A THEORY OF LEGAL SENTENCES

Law and Philosophy Library

VOLUME 34

Managing Editors

FRANCISCO J. LAPORTA, *Department of Law,*
Autonomous University of Madrid, Spain

ALEKSANDER PECZENIK, *Department of Law, University of Lund, Sweden*

FREDERICK SCHAUER, *John F. Kennedy School of Government,*
Harvard University, Cambridge, Mass., U.S.A.

Former Managing Editors
AULIS AARNIO, MICHAEL D. BAYLES†, CONRAD D. JOHNSON†, ALAN MABE

Editorial Advisory Board

AULIS AARNIO, *Research Institute for Social Sciences,*
University of Tampere, Finland
ZENON BANKOWSKY, *Centre for Criminology and the Social and Philosophical*
Study of Law, University of Edinburgh
PAOLO COMANDUCCI, *University of Genua, Italy*
ERNESTO GARZÓN VALDÉS, *Institut für Politikwissenschaft,*
Johannes Gutenberg Universität Mainz
JOHN KLEINIG, *Department of Law, Police Science and Criminal*
Justice Administration, John Jay College of Criminal Justice,
City University of New York
NEIL MacCORMICK, *Centre for Criminology and the Social and*
Philosophical Study of Law, Faculty of Law, University of Edinburgh
WOJCIECH SADURSKI, *Faculty of Law, University of Sydney*
ROBERT S. SUMMERS, *School of Law, Cornell University*
CARL WELLMAN, *Department of Philosophy, Washington University*

The titles published in this series are listed at the end of this volume.

MANUEL ATIENZA

and

JUAN RUIZ MANERO

Universiy of Alicante, Spain

A THEORY OF LEGAL SENTENCES

KLUWER ACADEMIC PUBLISHERS

DORDRECHT / BOSTON / LONDON

A C.I.P. Catalogue record for this book is available from the Library of Congress.

ISBN 0-7923-4856-7

Published by Kluwer Academic Publishers,
P.O. Box 17, 3300 AA Dordrecht, The Netherlands.

Sold and distributed in the U.S.A. and Canada
by Kluwer Academic Publishers,
101 Philip Drive, Norwell, MA 02061, U.S.A.

In all other countries, sold and distributed
by Kluwer Academic Publishers,
P.O. Box 322, 3300 AH Dordrecht, The Netherlands.

Printed on acid-free paper

English translation of Manuel Atienza and Juan Ruiz Manero,
Las piezas del Derecho, Teoría de los enunciados jurídicos,
Editorial Ariel, S.A., Barcelona, 1996.
Translator: Ruth Zimmerling

All Rights Reserved
©1998 Kluwer Academic Publishers
No part of the material protected by this copyright notice may be reproduced or
utilized in any form or by any means, electronic or mechanical,
including photocopying, recording or by any information storage and
retrieval system, without written permission from the copyright owner

Printed in the Netherlands.

TABLE OF CONTENTS

PREFACE ix

CHAPTER I: MANDATORY NORMS: PRINCIPLES AND RULES 1

1. Introduction. Types of principles 1

    1.1. The discussion about principles in contemporary legal theory: How it all started 1 / 1.2. Different meanings of 'legal principle' 3 / 1.3. A proposal of classification 5

2. Principles and rules 6

    2.1. A structural approach to the distinction 7 / 2.2. Principles and rules as reasons for action 12 / 2.3. Principles, rules, powers, and interests 15

3. The explanatory, the justificatory and the legitimatory dimension of principles 19

    3.1. Principles in legal explanations 20 / 3.2. Principles in legal reasoning 21 / 3.3. Principles, control and legitimation of power 25

APPENDIX TO CHAPTER I: REPLY TO OUR CRITICS 26

1. Mandatory rules as peremptory reasons, and principles as non-peremptory reasons; the 'closed' or 'open' configuration of the conditions of application 27

    1.1. Prieto's critique 27 / 1.2. Peczenik's critique 34

2. Principles and full compliance 36

    2.1. Prieto's position 37 / 2.2. Peczenik's position 41

CHAPTER II: POWER-CONFERRING RULES 44

1. Introduction 44

2. What power-conferring rules are not 45

    2.1. First exclusion: Power-conferring rules are not deontic or regulative norms 46 / 2.2. Second exclusion: Power-conferring rules cannot adequately be understood in terms of definitions, conceptual rules, or qualifying dispositions 53

3. What power-conferring rules are — 57

3.1. Three approaches and some ontological assumptions 57 / 3.2. A structural approach 58 / 3.3. A functional approach: Power-conferring rules as reasons for action 65 / 3.4. Power-conferring rules, non-normative powers, and interests 72

APPENDIX TO CHAPTER II: REPLY TO OUR CRITICS — 76

1. Introduction — 76
2. Critique of our critique of the deontic (or prescriptivist) conception — 76
3. Are we treating the conceptualist thesis fairly? — 84
4. Problems with our conception — 85

CHAPTER III: PERMISSIVE SENTENCES — 90

1. Permission in contemporary legal theory — 90

1.1. The pragmatic irrelevance of permissive norms. The category of 'permissive norms' is unnecessary. Echave-Urquijo-Guibourg (1980) and Ross (1968) 91 / 1.2. Von Wright's proposal: Permissive norms as promises 94 / 1.3. Weak and strong permission in Alchourrón and Bulygin 99

2. Reformulating the problem — 103

2.1. Permission and the regulation of 'natural' conduct 103 / 2.2. Permission and the exercise of normative powers 107 / 2.3. Permission and principles. Constitutional freedoms 112

3. Some conclusions — 115

APPENDIX TO CHAPTER III:
A NOTE ON CONSTITUTIONAL PERMISSION AND BASIC RIGHTS — 116

CHAPTER IV: VALUES IN THE LAW — 120

1. Introduction — 120
2. Two conceptions of criminal norms — 121
3. The double-faced character of norms and value judgments — 127
4. Types of norms and types of values — 135

Chapter V: The Rule of Recognition — 141

1. Introduction — 141
2. Jurists and the 'normative value' of the constitution — 142
3. The rule of recognition as ultimate norm — 145
4. Changing the rule of recognition without rupturing legal continuity? — 146
5. A host of problems — 147
6. Who shapes the rule of recognition? — 148
7. The conceptual, directive and evaluative dimensions of the rule of recognition. The rule of recognition and the exclusionary claim of the law. Why accept the rule of recognition? — 152
8. How many rules of recognition? Certainty and penumbra in the rule of recognition — 158

Chapter VI: Conclusions — 162

1. Introduction — 162
2. A classification of legal sentences: Table 1 — 162
3. A comparative analysis of the different types of sentences: Table 2 — 164

Table I. Classification of Sentences — 175

Table II. Comparative Analysis of Different Types of Sentence — 176

Bibliography — 183

Index of names — 191

Preface

This book is not a general theory of law, but it is intended to be the first part of one. It was written in the larger context of a rather comprehensive project we have been working on for a number of years and which we hope to finish some day. Our experience in working together makes us optimistic — maybe naively so — about our first results, presented here, as well as those we hope to obtain in the future, but not about the time this will take us. Let's say that, in this respect, the writing of the present book — this first chapter of a theory of law — has made us rather realistically pessimistic.

To our excuse, perhaps we can say that the topics treated here — and the same can be said of those we want to take on in the future — are extremely difficult. Proof of this is the fact that, although an enormous amount of intellectual work, often by very profound and original thinkers, has been spent on them, the basic problems are still far from being solved. Undeniably, in this century — especially since Kelsen — legal theorists have made essential contributions to the elucidation of the basic concepts of law. But it is no less true that this theory of law has been a domain visited only by specialists, and that, at least in countries like ours, it is still largely closed off from communication with academic and practicing lawyers who do not regard it as an adequate instrument for understanding, improving or criticizing the law, legal theory, or legal practice.

In our view, a theory of law intended to overcome this situation of self-absorption and isolation and to contribute to the dynamization of legal culture cannot confine itself to the use of the rich arsenals of analytical theory (whether in its version of a strictly logical analysis of the law or that based on the analysis of ordinary language). Besides, it must also incorporate two other great traditions of thought: a rationalist and objectivist conception of ethics, and a social philosophy that must have come to terms with Marxism.

The most important contributions to legal theory in this century have certainly come from the analytical philosophy of law; but that branch of philosophy is mainly a style of investigation, or a set of methodological tools, to be used in the service of ends other than its own development and refinement; that is, it should not be self-referential. Otherwise, when the tools (or means) are converted into ends — which happens rather frequently — there is the danger that the analysis will, at least in part, become an intellectual game blocking the investigation of substantive problems. And what should be kept in mind is the

— in our view — elementary, but often forgotten methodological rule that the aim should not only be to speak with rigour, but to speak, with the greatest possible rigour, about what really is important.

Although, as the reader will see throughout this book, we think that the identification of a norm as a legal norm is basically a question of social or institutional facts, and not one of a moral qualification, we do not think that it is possible, or that it would make sense, to elaborate a theory of law in total disregard of ethical theory. In this book, we expressly assume what could be called — though possibly in a minimal sense — a cognitivist and objectivist conception of ethics. We have various reasons for adopting cognitivism, but here we only want to mention one of them which is internal to the very intellectual task we are trying to carry out: If we would start from the position that moral judgments, in the last instance, are nothing but emotive expressions of approval or disapproval, or mere attempts to influence the behaviour of others, then we could not reconstruct fundamental aspects of our social practices, and especially of the legal practices typical of contemporary constitutional states. Besides, a minimally or moderately objectivist conception of ethics keeps, so to speak, an equal distance from both moral relativism and moral absolutism. Against relativism, we hold that moral judgments include a claim of correctness, and that to find out whether or not a moral judgment is correct is a matter of rational discussion. Against absolutism, we sustain that moral judgments — just like those of a court of last instance — express ultimate, but obviously not infallible reasons: any moral judgment (as, besides, any other judgment) is always open to scrutiny and rational criticism.

Finally, the political and social philosophy underlying our work, while — of course — not being Marxist, can be seen as the result of having come to terms with Marx — or a certain version of Marxism — which we have tried to lay down in another recent book.[1] Our contact with, or proximity to, that tradition of thinking has sharpened our awareness of the elements of inequality existing *even* in the law of contemporary developed societies, and of the ideological character of certain institutions — or aspects thereof — and legal concepts; it has also led us to acknowledge the historical character of law as well as of legal categories, and to try to develop a more 'materialistic' or social conception of law than that underlying the predominant theory — a conception, that is, which is not based exclusively on the notions of norm, legal relation-

---

[1] Manuel Atienza and Juan Ruiz Manero, Marxismo y filosofía del Derecho, Mexico City: Fontamara 1993.

ship, subjective law, etc., but which also takes into consideration other — in some sense, prior — notions like those of division of labour, human needs, interests, conflict, power, and so on.

Some readers will probably think that this is a 'declaration of principles' or a 'statement of purpose' that goes far beyond what can be found in the 'articles' that follow, i. e., in the six chapters the book consists of, which offer nothing but a theory of legal sentences, in the sense of a kind of taxonomy of the types of sentences[2] to be found in the law, understood as the language of the legislator. It is, one could say, the law seen as *langue*, i. e., as a set of rules for the use of legal language, and not as *parole*, i. e., as the different forms of usage of that (same) language by legal theorists, judges, lawyers, and others working in the field of law; because in judicial discourse or in legal dogmatics, there are sentences of a descriptive or explicatory kind that are not treated in this book. The hypothetical reader who would say this would, of course, be right; and we would even add that the ingredients we have put into our work are mainly — almost exclusively — products (the best, we think; in any case, of high quality) of the analytical theory of law. We do think, however, that our presentation — if you wish, a discussion with the highest authorities in analytical legal theory on what are, and how to understand, the linguistic units of the law: the legal pieces or *syntagms* — cannot be understood adequately if it is separated from the philosophical and methodological presuppositions and from the project of a general theory of law mentioned above. Thus, the theory of legal *sentences* will have to be followed by a theory of legal *acts*, and then a theory of legal *relationships* (the analysis of concepts like subject of law, subjective law, duty, responsibility, or sanction), and finally must end in a theory of legal *order* and *procedures* (of the creation, interpretation and application of law). The idea is to follow an outline that enables us to proceed from the parts to the whole, from a static to a dynamic perspective, and, if you wish, from the most simple and abstract to the most complex and concrete.

Legal sentences are, in our view, the most elementary units of law; but those pieces acquire full meaning only when their contribution to the shape and functioning of the law is well understood. They are not simply pieces of a puzzle, but of such a dynamic and tremendously complex reality as is the law of contemporary societies. A complete understanding of legal sentences, there-

---

[2] The expression 'sentence' is, of course, ambiguous: it can refer to a linguistic structure as well as to its meaning. In general, and unless we explicitly say otherwise, we will use the expression 'legal sentences' in order to refer to *meaningful* — i. e., interpreted — legal sentences.

fore, can be reached only in the context of a fully developed theory of law; thus, it will be a final result of that theory rather than a step in the construction. We have been well aware of that (necessary) constraint; but at the same time, we have tried to go beyond the mere structural analysis of legal sentences and to explore also what kind of reasons for action they provide and, thus, what role they play in legal reasoning. Finally, we tried — although this third perspective is less developed than the other two — to say something about the connexions between different types of legal sentences, on the one hand, and interests and power-relations between individuals and social groups, on the other.

Besides, although our analysis is confined to the level of legislative language, of the language of laws, we think that, in some sense, our results can be seen as a *general* theory of legal sentences: non-legislative sentences (where, obviously, the expression 'legislative' is used in its widest sense) either have no special characteristics that would justify a specifically legal analysis (this is the case of the descriptive and explicatory sentences mentioned above), or they can be seen as particular instances of legislative language, of the *langue* (a sentence, for example, would be a mandatory rule, but a particular, specific one, and also the result of the use of a power-conferring rule), or they could be analyzed as combinations of the types of sentences discussed here.

The typology of legal sentences we present starts with the distinction, within the class of mandatory norms, between rules and principles; rules as well as principles can, in turn, be either action norms or norms stipulating ends to be attained: all this will be treated in Chapter I. But legal orders also contain other kinds of sentences — power-conferring rules — which make it possible to introduce, modify, or derogate mandatory norms and, in general, to bring about normative changes. We propose to understand such sentences as anankastic constitutive norms. In their immediate neighbourhood, we find purely constitutive norms, more at a distance — since they are not norms — definitions. All these sentences are the subject of Chapter II. In Chapter III, we analyze permissive sentences and distinguish the context of rules regulating natural conduct from that of power-conferring rules and that of principles. Each one of the first three chapters is followed by an Appendix where we clarify or develop — in the first two cases, in a polemical form — some aspects already introduced in the main text. In Chapter IV, we consider evaluative sentences and value judgments contained in normative sentences. We then come to consider legal values as the justificatory aspect of norms understood as reasons for action. The rule of recognition is the topic of Chapter V: on the one hand, that ultimate rule provides a theoretical criterion for the identification of legal norms and, on the other, it has two practical dimensions: that of a guide for behaviour (especially the behaviour consisting in adopting legal decisions) and that of a criterion of evaluation (again, especially of those decisions). Finally, in the Conclusions we try to give an overview over it all, underscoring the similarities and differences between the different types of legal sentences. This is accompanied by two Tables which present a kind of synthesis of the book; they are a kind of 'map' of the types of legal sentences that may be useful to readers who, at any moment, overcome by the text's concentration on the exploration of the branches of some particular tree, wish to regain sight of the wood.

The process of grinding out this book has been arduous and has stretched over a long period of time. Preliminary, but not substantially different versions of the first two chapters were published as articles: a first version of ch. I in *Doxa* 10 (1991), a second version in *Analisi e diritto* (1993), and a first verison of ch. II also in *Analisi e Diritto* (1994). They have been criticized in the same journals by Luis Prieto and Aleksander Peczenik (*Doxa* 12 [1992]) and by Ricardo Caracciolo, Daniel Mendonça, José Juan Moreso and Pablo Navarro (*Analisi e diritto* [1995]). A first version of ch. III appeared in *Doxa* 15-16 (1994), of ch. V in the *Revista Española de Derecho Constitucional* 47 (1996). As readers will see, those critiques have not moved us to change the basic aspects of our conception; but they have helped us to fine-tune our analysis in several aspects, for which we are deeply grateful to all of them. We are also very much indebted to Francisco Laporta who over the past years has communicated to us numerous sharp observations that enabled us to improve our analysis in more than one point. Although Laporta is explicitly mentioned only in the Appendix to Chapter III, his presence is latent in many parts of the book. Our greatest debt is, of course, with our colleagues and friends from the seminar on legal philosophy at the University of Alicante (Josep Aguiló, Juan Antonio Pérez Lledó, Daniel González Lagier, Angeles Ródenas, Isabel Lifante, Pablo Larrañaga, Victoria Roca and Juan Antonio Cruz) with whom we have extensively discussed each chapter, in many, sometimes heated, sessions. Our very special thanks go to Daniel González Lagier whose profound knowledge of the work of G. H. von Wright — one of our main interlocutors in this book — has been of invaluable help, and to Josep Aguiló who, if not a co-author, must be considered at least a necessary accomplice of this work. The — constructively critical — contribution of them all has been so large that to dedicate the book to them seems to us the most natural way of expressing our gratitude.

CHAPTER I
MANDATORY NORMS: PRINCIPLES AND RULES

1. Introduction: Types of principles

1.1. The discussion about principles in contemporary legal theory: How it all started

The discussion about principles in recent legal theory started with a now famous article by Ronald Dworkin, published in 1967 under the title 'Is Law a System of Rules?' (later included as ch. 2 in Dworkin 1978). As most readers will know, the fundamental purpose of that article was what Dworkin himself called an "attack" on legal positivism, and especially on the "powerful form" it has found in H. L. A. Hart's legal theory. According to Dworkin, one of the main drawbacks of that theory is that it cannot account for the presence in the law of standards other than rules, namely, of principles. This also makes it impossible for Hart's construction to comprehend essential aspects of legal reasoning in so-called hard cases. According to Dworkin, rules and principles differ in the following way:

"Both sets of standards point to particular decisions about legal obligation in particular circumstances, but they differ in the character of the direction they give. Rules are applicable in an all-or-nothing fashion. If the facts a rule stipulates are given, then either the rule is valid, in which case the answer it supplies must be accepted, or it is not, in which case it contributes nothing to the decision [...] But this is not the way [...] principles [...] work. Even those which look most like rules do not set out legal consequences that follow automatically when the conditions provided are met [...] Principles have a dimension that rules do not — the dimension of weight or importance. When principles intersect [...] one who must resolve the conflict has to take into account the relative weight of each [...] Rules do not have this dimension." (Dworkin 1978, 24 ff.)

As an example for his thesis, Dworkin invokes a case — *Riggs v. Palmer* — that had to be decided by a New York court towards the end of the last century. The facts were basically the following: a grandson requested to be given possession of his grandfather's inheritance which, according to New York inheritance law, belonged to him. The peculiarity of the case was that among the attributes of that grandson was that of being the murderer of the testator — a circumstance not mentioned as a reason for exclusion from succession by the applicable inheritance law. Thus, according to the applicable legal statute that circumstance had to be regarded as irrelevant. The court, however, denied the grandson his grandfather's inheritance, on the principle that "[n]o one shall be permitted to [...] take advantage of his own wrong". It may be observed that in at least two points the example used by Dworkin to defend his theses is not quite consistent with those theses:

    a) According to Dworkin, if a rule is valid, and if the facts it stipulates are given, then the application of the rule directly solves the case. But in *Riggs v. Palmer* the court did not question the validity of the applicable provisions on inheritance; rather, it held that with respect to the circumstances of the case at hand, the principle of *Nemine dolus suus prodesse debet* should prevail over the principle that what is prescribed by legal rules ought to be done.

    b) According to Dworkin, principles alone never completely determine the content of a particular decision; but in the case he invokes, it seems that, once the court determined the prevalence of the principle prohibiting to take advantage of one's own crime, that principle became the only and complete foundation of the decision.

However this may be, Dworkin's theses on principles, later expanded into the global conception of law to be found in *Law's Empire* (Dworkin 1986), have been among the most significant stimulants of the discussion in legal theory and philosophy of recent years, leaving their marks on such central topics of those disciplines as the relationship between morality and law, the typology of legal provisions, the characterization of hard cases, the judicial creation of law, the character and structure of justificatory legal reasoning, etc. Our purpose in this chapter, however, is not to report on and participate in that debate, or to show our agreement or disagreement with its main protagonists. Although we will freely use some of the elements brought up in the course of that debate, our aim is to construct a first step of a theory of legal sentences, using as our starting point the distinction between rules and principles (and between all the other subspecies we think should be distinguished within each one of the two species). In our view, with respect to each and every type of sentence, such a theory should clarify the following: First of all, what are its structural characteristics? Second, what kind of reason for action does it provide? And third, how does it relate to the interests and power relations existing in a society?

1.2. Different meanings of 'legal principle'

It should not be forgotten that, although the discussion on legal principles in contemporary legal theory has its origin in Dworkin's work (and to a great extent still concentrates on it), legal theory before Dworkin has not been totally void of reflections on legal principles. Works like those of Esser (1956), Del Vecchio (1958), Bobbio (1966), or García de Enterría (1963) in Spain, are excellent proof of this. And as for statutory texts, there is a long tradition of reference to legal principles, usually traced back to the Austrian Civil Code of 1811. However, in legal theory and practice the expression 'legal principles' (or 'general principles of law') has been — and still is — used with different, only partly overlapping meanings. The most common and significant ones seem to be the following (here, we basically follow the analysis of Carrió [1986] and Guastini [1990]):

a) 'Principle' in the sense of a very general norm, understood as a norm regulating a case whose relevant properties are very general. For example, art. 1091 of the Spanish Civil Code, which stipulates that "Obligations arising from contracts have legal force between the contracting parties and must be complied with as contracted". It hardly needs to be said that the property of being general is a relative matter and can come in degrees: the norm (or principle) just mentioned is more general than those applying exclusively to contracts of lease (e. g., art. 1545 of the Spanish Civil Code: "Commodities which are consumed in their use cannot be the object of such a contract"), and less general than those applying to contracts as well as other acts-in-the-law (for instance, art. 11.1 of the Spanish Civil Code: "The forms and formalities of contracts, wills, and other acts-in-the-law shall be regulated by the law of the country in which they are celebrated"). In the sense in which the expression is employed here, the generality of a norm is not a quality referring to whether the class of its addressees is small or large; rather, it refers to the greater or lesser generality (or 'genericity', if you like) of the relevant properties of the case it regulates.

b) 'Principle' in the sense of a norm clad in particularly vague terms — like art. 7, para. 2 of the Spanish Civil Code: "The law does not protect the abuse or anti-social exercise of a right". There are, of course, many norms that are vague in the sense that in the description of the generic case (or legal facts) we find terms with an open-textured periphery: Most ordinary cases clearly fall inside or outside of a norm, but there are also (exceptional) situations where it is doubtful whether or not a norm applies. The example given, however, points to another kind of vagueness that is produced when one uses what legal theorists call *indeterminate legal concepts*, i. e., terms (such as "abuse of a right") which not only have an open-textured periphery but also a vague core. Here too, we can find cases clearly covered by, or clearly falling outside of, the norm in question; but the great majority of real cases requires specification, that is, a weighing of relevant factors whose identity and possible combinations cannot be anticipated.

c) 'Principle' in the sense of a program norm or policy, that is, a norm stipulating the obligation of pursuing certain ends. For example, art. 51.1 of the Spanish Constitution: "The powers of the state shall guarantee the defence of consumers and users, protecting their security, health and legitimate economic interests through effective procedures".

d) 'Principle' in the sense of a norm expressing the highest values of a legal order (reflecting a certain way of life), or of a part of it, of an institution, etc. For example, art. 14 of the Spanish Constitution: "All Spaniards are equal before the law; there shall be no discrimination on grounds of birth, race, sex, religion, opinion or any other condition or personal or social circumstance".

e) 'Principle' in the sense of a norm directed at the organs entrusted with the application of the law, and indicating how the applicable norm must generally be selected, interpreted, etc. An example of such a principle could be art. 4, para. 2 of the Spanish Civil Code: "Criminal laws, laws of exception and temporally restricted laws shall not be applied to other facts, nor at other times, than those explicitly stated in them".

f) 'Principle' in the sense of *regula juris*, that is, a sentence or maxim of legal science with a high degree of generality, permitting the systematization of the legal order or a sector thereof. Such principles may or may not be incorporated into positive law. An example of the former (at least in Spanish law) is the principle of normative hierarchy (art. 9.3 of the Spanish Constitution); and of the latter, the principle of the rational legislator which, though not explicitly formulated in our law, is often used by academic and practicing lawyers, e. g., in order to sustain that a norm ought to be interpreted in some specific way (alleging that only if it is interpreted in this way one could say that the legislator followed some rational purpose in issuing that norm).

1.3. A proposal of classification

This list of meanings is by no means exhaustive (without claiming to exhaust the senses in which legal theorists speak of 'principles', Carrió distinguishes eleven different concepts), nor is it exclusive: the meanings overlap, and as the reader probably noticed, a good number of the norms used as an example for one sense of 'principle' could have been used also as an example for other senses of the expression. In order to avoid as much as possible the imprecisions inevitably connected with the use of such an ambiguous and vague term, we will begin with the following triple classification of principles.

1.3.1. First of all, we can distinguish between *principles in the strict sense* ['principles' in the sense of *d)*] and *policies or program norms* ['principles' in the sense of *c)*]. This distinction we understand to be exhaustive and exclusive. To regard the distinction as exhaustive implies that properties *a)* — great generality — and *b)* — presence of centrally vague terms — which normally accompany both principles in the strict sense and policies do not by themselves allow us to qualify some standard as a principle. As for *e)* and *f)*, standards with these properties, if they also have properties *a)* and/or *b)*, can be reduced either to principles in the strict sense or to policies. To see the distinction as exclusive implies that, although one and the same sentence can be understood as a principle in some contexts of reasoning and as a policy in others (and it can even be said that this is an ambiguity that is typical of many principles), one and the same person cannot use them, in one and the same context of reasoning, in both senses at the same time.

1.3.2. The second distinction we think is important is the one that can be drawn (freely using Alchourrón and Bulygin's [1971] terminology, with a scope somewhat different from theirs) between *principles in the context of the primary system* or *system of the subject*, and *principles in the context of the secondary system* or *system of the judge* (and, generally, *of legal organs*), that is, between principles (standards of conduct that can be formulated as principles in the strict sense or as program norms) insofar as they are intended to guide the conduct of the general public, and principles insofar as they are intended to guide the exercise of public normative powers (the creation and application of norms) by legal organs. This distinction is exhaustive, but not exclusive. There are, of course, principles guiding only the exercise of public normative powers; but there are none that operate only on the conduct of general subjects: all principles of which it can be said that they guide the conduct of the addressees of legal norms in general at least constitute a criterion for the evaluation of that conduct by law-applying organs.

1.3.3. Finally, the last distinction we are interested in is that between *explicit principles*, i. e., principles explicitly formulated in a legal order, and *implicit principles*, i. e., principles derived from other sentences existing in a legal order (e. g., the principle that norms are to be interpreted as if they had been issued by a rational legislator). Obviously, this distinction is exhaustive and exclusive.

2. Principles and rules

Besides these *internal* distinctions, principles must also be distinguished, so to speak, *externally*, from other standards of behaviour that can be part of a legal system. Here, we assume that legal systems consist not only of mandatory norms, but also of other sentences (like permissive provisions, definitions and power-conferring rules, to be considered in the following chapters), and that mandatory norms can, in turn, be *rules* or *principles*.

The problem, therefore, can be formulated as follows: How can principles be distinguished from rules? And, at the same time, what is the scope of the classifications of principles previously mentioned? Our strategy in answering these questions will be to start from three typical approaches often adopted in dealing with norms and to apply them to the problem we are interested in.

The first one we can call the *structural* approach, because it consists in regarding norms as entities organized in a certain way. An example of this is the conception of norms to be found in Alchourrón and Bulygin's *Normative Systems* (1971), where norms are conceived as sentences correlating generic cases (sets of properties) to solutions (that is, to the normative qualification of certain conducts).

The second typical way of understanding norms could be called *functional*, since it concentrates on the role or function they — claim to — have in the practical reasoning of their addressees. An example of this approach in contemporary legal theory would be especially the work of Joseph Raz, and one of H. L. A. Hart's later articles. As you may know, in those works norms are regarded as reasons for action; from this perspective, the fundamental interest is to show what kind of reasons norms are, and how they operate in practical reasoning.

The third approach is to look at norms not in terms of cases and solutions, or of reasons for action and practical reasoning, but in connection with the interests and power relations existing in a society. On the one hand, legal norms can be seen as the result or effect of certain social relations; on the other, legal norms themselves shape the relations between subjects by conferring powers and protecting the interests of some against others; and, finally, the power given to some individuals (or groups) by legal norms is used to change — or, more generally, to affect — social relations themselves. This third way of approaching norms — more closely linked to the sociology of law or the ideological critique of the law — is usually missing in the main currents of contemporary legal theory. In our view, that is a substantial deficit. As we will try to show later, taking into account this third perspective too can contribute enormously to a clarification of the role of principles in legal life.

Depending on which perspective one chooses for one's conception of norms, the question of how to distinguish principles from rules, or different kinds of principles from each other, can take on very different tones.

2.1. A structural approach to the distinction

For the first approach, the question is whether principles, like rules, can always be formulated in a conditional pattern, that is, one that correlates cases and solutions (this does not imply to accept — nor to reject — that norms are linguistic entities; it only implies acceptance of the fact that they can be expressed in some language). Before we answer this question, we must clarify a few things. The first is that the pattern used by Alchourrón and Bulygin, and which we have taken as our starting point, understands legal norms as correlations between generic cases (sets of properties) and solutions (that is, the normative qualification of certain behaviours). It must be noted, however, that as far as *rules* are concerned, to see the consequent (the solution) of norms in terms of a normative qualificaton *of a certain behaviour* is adequate only for the most common type of legal rules, which we propose to call *action rules*. Besides such rules, legal systems also contain rules which deontically qualify the *attainment of some state of affairs*, rather than a behaviour. We propose to call mandatory rules of this last kind *end rules*. Sometimes, the distinction between action rules and end rules is important, while on other occasions it may be merely a question of legislative style. Thus, for example, it makes no difference whether one prohibits — on the condition, of course, that there is no reason of justification — the action of killing or the state of affairs consisting in having killed someone. In that sense, any rule formulated in terms of an action rule can be 'translated' into the terms of an end rule (where the two are only two different ways of saying the *same* thing). But the distinction becomes relevant when the provision stipulating as obligatory, for example, the production of a certain state of affairs leaves the selection of the *causally appropriate* means for bringing it about to the discretion of its addressee. In that sense, end rules allow their addressees a *margin of discretion* that does not exist in the case of action rules. As an example for an end rule we can use article 103, 3 SCC. That provision says that whenever a petition of nullity, separation, or divorce has been accepted and there is no agreement between the parties, the judge must, among other things, adopt the following measures: "Determine the contribution of each spouse to the burdens of the marriage [...]; establish the bases for calculating the quantities, and *fix guarantees, deposits, reserves or other convenient measures of precaution, in order to secure the effectiveness of the respective sums one of the spouses has to pay to the other*". What the judge is being ordered to do here is to bring about a certain state of affairs (that in which "the effectiveness of the respective sums one of the spouses has to pay to the other" is guaranteed), whereas she has a margin of discretion when it comes to choosing the appropriate means for bringing this about ("fix guarantees, deposits, reserves or other convenient measures of precaution").

The second clarification refers to which of the previous classifications of principles are relevant to the effects we are interested in now. It seems obvious that it is not the distinction between explicit and implicit principles, nor that between principles in the context of the primary systems and principles in the context of the secondary system. If there were some structural difference depending on whether the case regulated by a norm consists in, or refers to, the exercise of a public normative power, it would be logical to think that it would apply to principles as well as to rules. So, what we must look at is whether or not principles in the strict sense and policies have a conditional structure (that is, whether they correlate cases and normative solutions) and/ or whether the deontic operators applying to rules are the same as those applying to principles.

2.1.1. In our opinion, principles in the strict sense can always be formulated as sentences correlating cases to the normative qualification of a certain behaviour; but that does not mean that from this perspective there is no difference to rules (and especially to action rules). The difference is that principles present the case in open form, whereas rules present it in closed form. By this we mean that, while in rules the properties constituting the case are a finite and closed set, in principles no closed list of properties can be formulated. Not only do the properties constituting the conditions of application have a larger or smaller periphery of vagueness; those conditions are not even generically determined. Thus, the kind of indeterminacy principles suffer from is more radical than that of rules (although, of course, between the two kinds of indeterminacies there can be an area of penumbra).

Robert Alexy, in developing a point which already seems to be present in Dworkin, has written that

"the decisive point for the distinction between rules and principles is that principles are norms ordering that something be implemented to the highest possible degree, relative to what is legally and materially possible. Thus, principles are mandates of optimization whose characteristic property is that they allow for different degrees of compliance, and that the amount of compliance they command does not depend only on the material, but also on the legal possibilities given in each case. The extent of legal possibilities is determined by principles and rules pulling into opposite directions. Rules, in contrast, are norms requiring full compliance, and insofar they can always be only complied with or violated. If a rule is valid, then it is obligatory to do exactly what it prescribes, no more and no less" (Alexy 1988, 143 f.; see also Alexy 1986, 75 f.).

In our view, it is true that principles can be complied with in different degrees, as far as policies or program norms are concerned, but not in the case of principles in the strict sense. Let us elucidate this with the help of a few examples.

A principle like the one formulated in art. 14 of the Spanish Constitution, understood as a secondary principle, can be presented, we think, in the form of a conditional like the following: "*If* (condition of application) a legal organ uses its normative powers (that is, issues a norm in order to regulate a generic case, or applies a norm in order to solve an individual case, etc.), and with respect to the individual or generic case in question there is an opportunity for discrimination on grounds of birth, race, sex, religion, opinion or any other personal or social circumstance, and there is no countervailing principle of higher weight in the case at hand, *then* (normative solution) the organ is prohibited to discriminate on the grounds of any of the factors just mentioned."[1]

---

[1] Principles could be said to be 'categorical norms' in the sense of von Wright, that is, norms whose "condition of application is the condition which must be satisfied if there is going to be an opportunity for doing the thing which is its content, *and no further condition*" (von Wright 1963, 74). Art. 14 of the Spanish Constitution prohibits *prima facie* to discriminate on grounds of birth, race, sex, etc., whenever there is an opportunity for discriminating on such grounds; that *prima*

Here, we find the typical indeterminacy of principles only in the openness of the conditions of application, not in the description of the behaviour that is prohibited, i. e., that of discriminating. In some contexts, 'discrimination' can, of course, be understood as a vague term, but this kind of vagueness also exists in the standards we call 'rules'. The rule that a working woman should receive the same pay as a man differs from the above principle only in that its conditions of application are not open (thus, art. 28 of the Spanish Workers' Statute stipulates that "Employers must pay an equal salary — with respect to the basic salary as well as to salary supplements — for equal work, without any discrimination on grounds of sex"); but even here, there may be a problem of vagueness when it comes to determining whether in a particular case the conditions of application obtain (there may, e. g., be doubts about whether a certain activity should be considered 'work') or what the description of the prohibited behaviour amounts to (should different salary supplements for clothes on grounds of sex be considered prohibited by art. 28?). Structurally, the only difference between art. 14 of the Constitution and art. 28 of the Workers' Statute is that in the latter case the norm's conditions of application form a closed set, although its formulation may be plagued by problems of semantic indeterminacies (we insist that semantic indeterminacies do not affect the character of a rule, unless they reach such an extreme degree that one cannot even say that they determine the conditions of application). But concerning the description of the deontically qualified pattern of behaviour, the two norms display a similar degree of determinacy. In other words: once it is clear that, for the combination of relevant factors in some particular case, the principle of equality sanctioned in art. 14 has prevalence over countervailing principles or rules, that principle

---

facie prohibition gives way only if with respect to the case at hand another principle applies which counteracts the former and which, in the case in question, has higher weight (or, if you like, if the reasons in favor of unequal treatment deriving from some other principle have more weight than the reason given by the principle of equal treatment).

Rules, in contrast, are intended to preclude deliberation; by correlating their normative solution with conditions of application (or generic cases) consisting in properties that are independent of the reasons speaking for or against that normative solution, they *claim* to impose obligations or prohibitions not merely *prima facie*, but *all things considered*, for all cases where those conditions of application obtain (or which can be subsumed under the respective generic case). However, as we will see later, the scope of this claim seems to be restricted by the possibility that in some cases, the application of a rule may come into conflict with a principle that, with respect to the relevant properties of the case, has greater weight than the principle(s) sustaining that rule. On this question, see infra, ch. I, 3.2.2.1., and the Appendix to ch. I, 1.1., point 4.

requires full compliance: it is either obeyed or not obeyed; and there is no way of compliance by degree.

2.1.2. When we now pass from principles in the strict sense to policies or program norms, things seem to look somewhat different. The distinctive character of policies is that this kind of standards *leave their conditions of application as well as the prescribed pattern of behaviour open*. As an example for a policy, we will use art. 51.1 of the Spanish Constitution, already mentioned earlier. As in art. 14, here the conditions of application are formulated in an open fashion: the provision does not state under what conditions state agencies must act as prescribed, that is, protect consumers and users. But in contrast to art. 14, art. 51 also does not command — nor prohibit — any action; instead, it commands that a purpose, a state of affairs with certain characteristics be attained, namely, that the security, health and legitimate economic interests of consumers and users are effectively protected. What actions (or courses of action) are causally appropriate for reaching this goal is not determined in the Constitution. And the objective attainment of which is commanded — that the security, health and legitimate economic interests of consumers and users are effectively protected — itself obviously is not the only goal commanded in the Constitution.

Here, we must underscore two things: on the one hand, that courses of action appropriate for attaining some constitutionally commanded purpose (e. g., making mortgages cheaper, or liberalizing the use of land, as appropriate means for making it easier for all citizens to provide decent and adequate housing for themselves — an objective stipulated by art. 47 of the Spanish Constitution) to the highest possible degree can have negative effects on other objectives also constitutionally commanded (e. g., on economic stability — art. 40 of the Spanish Constitution — in the first case, or on an "environment adequate for the development of people" — art. 45 of the Spanish Constitution — in the second); and on the other, that in many cases such constitutionally commanded objectives are interdependent: for example, the weakening of economic stability can have negative effects on access to housing or on social security systems. Thus, in contrast to what happens with principles in the strict sense, in behaviour directed by policies the question is not which one prevails over the other in some particular case, but to formulate policies capable of attaining, to the highest possible degree, the joint realization of all objectives. For what has been said above, in our view, Robert Alexy's theory of principles as *mandates of optimization* distorts the matter with respect to legal principles in the strict sense, but it seems perfectly adequate to account for policies.

Among principles (in the wide sense), policies are the *pendant* of end rules. The difference between end rules and policies is that the former state their conditions of application in a closed way, whereas the latter leave them open, and especially that the former stipulate an end that must be attained fully, and not only, as is the case with the latter, to the greatest possible extent, taking into account the existence of other ends and the available means.

2.2. Principles and rules as reasons for action

The second way of understanding norms — as reasons for action — is probably more illuminating than the previous one when it comes to principles. In order to make analysis easier, we will begin by considering only principles in the context of the secondary system, that is, principles as standards directed at norm-authorities and, more specifically, at judicial organs, understood in a wide sense (what Raz calls 'primary organs'), i. e., organs to which the law itself confers the *normative power* to *authoritatively* solve disputes, and on which it imposes the duty to do this *legally* (i. e., grounding their decisions on the standards identified as legal). For the time being, we will also ignore the distinction between principles in the strict sense and policies.

2.2.1. Now, in this context, the distinction — and the relationships — between rules, explicit principles and implicit principles can be better understood if seen from the perspective of the distinction — and the relationships — between obedience to norm-authorities and one's own deliberation, with respect to the law-guided behaviour of judicial organs.

2.2.1.1. We will start from a well-known characterization by the later Hart (1982a) and regard *legal rules of action* as peremptory and content-independent reasons for action. 'Peremptory reason' means the same as 'protected reason' (in Raz's terminology), that is, a first-order reason for performing the required action — in the context we are interested in: for issuing a decision whose content is in accordance with the rule — and a second-order reason "to preclude or cut off any independent deliberation by the hearer of the merits pro and con of doing the act" — that is, in our context, to preclude that the content of the decision will be grounded on what the judicial organ, in view of the merits of the case, thinks would be the best solution. Thus, with rules it is intended that, when their conditions of application obtain, judicial organs suspend their own judgment about the balance of applicable reasons as the basis of their decision and, instead, adopt the content of the rule. That they are content-independent refers to why judicial organs should obey the rules, that is, to the reasons why they should regard them as 'peremptory' or 'protected'. It is assumed that judicial organs should do this because of their source of origin (that is, the norm-authority who has issued or adopted them). Their origin in that source is the reason why judicial organs should regard them as peremptory reasons.

End rules too are peremptory and content-independent reasons for bringing about the prescribed state of affairs. However, they show important differences, as compared with action rules, in how they guide behaviour. Action rules simplify the decision process of those who should behave according to them (i. e., those who should comply, or control compliance, with them). They only need to determine whether or not certain conditions hold, in order to perform or abstain from performing a certain action, irrespective of the consequences, that is, of the causal process their behaviour will set in motion. End rules, in contrast, delegate the control of (or responsibility for) the consequences of their behaviour to the addressees themselves.[2]

2.2.1.2. Of *explicit principles* we can say that they are content-independent, but not peremptory reasons for action. They are content-independent because the reason why they are reasons for the action of judicial organs, that is, the reason why they should be part of the justificatory reasoning for the decisions of judicial organs, is the same as in the case of rules, namely, that they have their origin in a certain source. But they are not peremptory, because they are not intended to preclude deliberation by the competent judicial organ about the content of the decision to be issued; rather, they are merely first-order reasons for deciding in a certain sense whose force as compared to that of other reasons

---

[2] On the other hand, it seems clear that the conduct of judicial organs as such in general is not guided by end norms, but by action norms (although there are examples — like the one mentioned before — of end rules addressed to a judicial organ as such). But in any case, judicial organs should control the compliance with end rules by their addressees, among them especially administrative organs. For that control, they must take into account the regulated character of the state of affairs appearing in the consequent of the end rule, as well as the discretional character of the choice of the means to reach that state of affairs. As readers may have noticed, the distinction between action rules and end rules as it has been sketched here is indebted to Niklas Luhmann's distinction between the "conditional" and the "purposive" programming of decisions (cf. Luhmann 1974). An application to the problem of administrative discretion (and its judicial control) can be found in Atienza (1995).

(other principles) — which may, in turn, be reasons for deciding in another sense — must be weighed by the judicial organ itself.

2.2.1.3. Finally, *implicit principles* are reasons for action that are neither peremptory nor content-independent. They are not peremptory for exactly the same reason as explicit principles; and they are not content-independent because the reason why they should be part of the reasoning of judicial organs is not their origin in some specific source, but some quality of their content. In our opinion, that quality is not their intrinsic justice (i. e., principles are not directly derived from social morality, as Dworkin sometimes seems to believe [1978 and 1986], nor are they simply deep-rooted social standards, as some experts in civil law — e. g., Díez Picazo [cf. Díez Picazo/Gullón 1989] — hold); rather, they fit, or are coherent with, those rules and principles based on sources.

Thus, we can say that rules are authoritative reasons as much with respect to *why* they operate as with respect to *how* they operate in the justificatory reasoning of judicial organs; that explicit principles are authoritative reasons with respect to why they appear in such reasoning, but not with respect to how they operate in it; and that implicit principles are not authoritative with respect to their way of operating in such reasoning, and that as far as the reason for their presence is concerned, they are authoritative at best indirectly, insofar as their presence is based on their fit to rules and explicit principles.

2.2.2. When we now go from secondary principles to primary principles, that is, principles as standards addressed to the general public, there is no reason why the analysis should be very different from the former. The difference here seems to be rather that, in contrast to judicial organs, people in general — except in very special cases — do not need to justify their behaviour with regard to legal norms. But besides that, for those who accept them, rules operate as peremptory or protected reasons, whereas principles operate as first-order reasons that must be weighed against other reasons.

2.2.3. With respect to the distinction between principles in the strict sense and policies, we think that in terms of reasons for action the difference can be sketched as follows. Policies generate reasons for action of a utilitarian kind: The fact that it is desirable to attain some end E implies that, in principle, there is a reason in favor of whatever leads to that end; but the reason is not exclusionary, since there may be other, countervailing reasons of greater force. In contrast, reasons for action deriving from principles in the strict sense are rightness reasons: like the former, they are not exclusionary; but in a subject's deliberation, rightness reasons operate as ultimate reasons. Hence, utilitarian reasons deriving from policies can and should be evaluated (and, if appropriate, overruled) by rightness reasons based on principles, whereas the contrary cannot happen: if one has a rightness reason to do X, then not to do X can be justified only by invoking other reasons of the same kind (i. e, principle-based reasons) with higher weight, but not by invoking (policy-based) utilitarian reasons showing that to attain a certain end is incompatible with action X.

On the other hand, by stipulating that performance of a certain behaviour is obligatory (or prohibited) in their consequent, principles in the strict sense allow their addressees to disregard the consequences of their actions. That means that once a principle has been determined as prevalent in a specific case, its addressee is in the situation of the addressee of an action rule: he should simply do what is prescribed, without having to pay any attention to the causal process the performance of that conduct will start. Policies, on the contrary, require their addressees to deliberate on the adequacy of their behaviour (that is, of the means they employ) in view of the pursued end as well as in view of other ends whose pursuit also has been ordered and which may be negatively affected by the use of those means.[3]

2.3.   Principles, rules, powers, and interests

In order to explain how the concepts of power and interest can help elucidate the concept of legal principles, it may be useful to begin by looking at those two rather problematic concepts.

2.3.1. With good reason, the concept of 'power' has been considered a paradigmatic example of an 'essentially contested' concept. However, as Steven Lukes has shown by introducing a distinction similar to that of Rawls (1971) between the concept and conceptions of justice, it is not so much the concept that is

---

[3] This has consequences for the judicial control of compliance with principles in the strict sense or with policies. In the first case, in order to evaluate whether an addressee's conduct is lawful or unlawful, the judicial organ only needs to determine whether in the case at hand the principle in question has prevalence. In the second, however, it must take into account that the addressee — usually, some legislative or administrative organ — enjoys discretion with respect to the choice of the appropriate means for jointly satisfying to the highest possible degree all the different ends it has been ordered to maximize.

contested as the diverse conceptions of, or approaches to, power. Thus, one can formulate a concept of power underlying all or at least most conceptions of power to be found in the social sciences. According to Lukes, that concept underlying the diverse conceptions of power can be defined as follows: "A exercises power over B when A affects B in a manner contrary to B's interests" (Lukes 1974, 27).

Now, for two reasons, that concept is still too narrow for our purpose: first of all, because we are not only interested in the *exercise of power*, but also in *having power* as such; and secondly, because we are interested in a notion of power that does not exclude cases where the interests of others are not affected *negatively*. Therefore, our definition will be this one: "A has power over B when A has the ability to affect B's interests". That definition — or, if you wish, that concept of power — can be interpreted in different ways: it can give rise to different conceptions of power. Of these, the most general — and the one we will adopt — is characterized by the understanding that *1)* A and B can be individuals as well as groups, i. e., social classes, pressure groups, and so on; *2)* in order to 'have the ability' it suffices, on the one hand, that B believes A to have it, although in fact this is not so, and, on the other, A can have the ability without knowing or being aware of it; *3)* A's ability to affect B's interests may be either negative (that is, the power to harm B's interests) or positive (a power exercised to the benefit of someone else — e. g., a father's power over a child, provided it is used correctly — also is power), and that *4)* this may concern subjective as well as real or objective interests of B. By the latter, we understand those interests B is not aware of, but which it is reasonable to assume she would perceive to be her interests — on the basis of what B herself considers to be her ultimate interests — if she did possess the relevant information about the pertinent causal relations.[4] What we have now is a concept of power that is very wide but which connects that notion in an essential way to that of interest: power is a kind of relation where the participating subjects are in a situation of inequality in the sense that some can affect the interests of others.

---

[4] As can be seen, we use the expression 'objective interests' in what could be called a minimal sense: we do not refer to something an agent simply should be interested in, but to something he should be interested in — although he actually isn't — *in the light of his ultimate subjective interests*. For a critique of other, stronger concepts of the term 'objective interests' cf. Bayón 1991a, 114 ff. In general, with respect to the notion of 'interest' and, later, also of 'value', we somewhat freely follow Bayón.

2.3.2. Generally, there are a number of different ways in which legal norms are linked to interests and power. As has already been said, on the one hand, they are the effect of interests and power relations; on the other, they legally shape power relations; and finally, the exercise of those powers has the effect of bringing about changes in the power relations and interests existing in a society. Right now, we are interested in the second of those linkages between norms and power. Here, what is important is that, with respect to legal norms, not only does power appear at the time of their stipulation or application: legal norms themselves also shape a power structure, i. e., they confer on certain individuals or groups the ability to affect the interests of other individuals or groups.

A minimal justification of the existence of a legal order as such lies in the fact that individuals and groups have interests whose mutual reconciliation cannot be expected to arise spontaneously in social life. In other words, without the existence of legal norms social life would be impossible, or at least extremely costly. Now, that function of the reconciliation — or normativization — of interests can obviously be fulfilled in different ways.

2.3.2.1. One of them is to proceed through legal provisions enabling their addressees to develop their plans of life without having to think about how their actions may affect the interests of other social subjects in each single case. This is typical of those legal norms we have called 'action rules', and especially those belonging to what, following Alchourrón and Bulygin, we have called the 'system of the subject' (and which, besides, according to the 'system of the judge', form part of the criterion on which judges must base their evaluations of the conduct of 'subjects'). Such rules make constant weighing and deliberating unnecessary, by imposing restrictions on everyone's pursuit of his own interests (rules imposing positive or negative duties) and by guaranteeing a sphere of non-interference by other social subjects for that pursuit (permissive rules indirectly imposing prohibitions of interference on others).

2.3.2.2. This way of operating, however, is inadequate when one sees the law — norms — as having not only the function of delimiting the area within which everyone may pursue his or her own interests, but also that of actively promoting certain social interests. For this purpose, action rules are not sufficient; end rules and policies must be stipulated, and they typically treat interests in a different way. But the difference is not that this second kind of norms, in contrast to the first, does not suppose the existence of conflicts of interests (neither end rules nor policies are technical norms presupposing that ends are given, such that the only problem, therefore, is that of the best means to reach those ends). The difference, rather, is that end rules, and especially policies, do not *ex ante* limit the articulation of conflicting interests — nor, in the case of policies, of at least relatively incompatible objectives; rather, in every single case they require deliberation on those interests in order to establish their relative weights. Policies, we can say, do not determine spheres of power once and for all, irrespective of the interests actually present in each particular case — as action rules do. They make that determination depend on circumstances that are variable and cannot be determined *a priori*, that is, circumstances not contained in the norms.

2.3.2.3. Finally, legal orders impose restrictions on the pursuit of interests by the different social subjects by incorporating values regarded as *categorical reasons with respect to any interest*. That is why — as has already been said — the norms transporting those values — i. e., principles in the strict sense — have precedence over policies and mainly play a negative role: principles in the strict sense neither intend to regulate the clash of interests, nor to promote any specific interest; rather, they try to prevent that the pursuit of any interest may harm those values. Obviously, that those values are considered categorical reasons against all interests does not preclude that they may themselves come into conflict. At least liberal-democratic legal orders contain a host of values whose internal rank order is not always predetermined by the legal order itself. There is, thus, a possibility of conflict that can only be resolved by deliberation on which value, in view of the circumstances of the case, has the higher weight.

2.3.3. What has just been said should suffice to elucidate, from that third perspective of norms, the difference between rules and principles as well as between the two kinds of principles distinguished above, i. e., policies and principles in the strict sense. Now, we must ask whether this kind of analysis can also be used with the other distinctions made earlier, concerning legal principles.

2.3.3.1. As for the distinction between explicit and implicit principles, possibly the most interesting question to consider from this perspective is whether implicit principles (regardless of whether they are policies or principles in the strict sense) are connected with the hidden prevalence of certain interests and values in a legal order, or with the equally hidden incorporation of certain power relations in it (irrespective of whether those who issue or accept the norms are aware of it). In our view, such a connection exists sometimes, but not always. In other words, that the principles in a legal system are not always clearly stated is due, in part, to technical reasons arising from the very nature of principles: on the one hand, social dynamics constantly cause new objectives and purposes to arise that could not be foreseen by the norm-creating organs; on the other, principles depend on rules and are, at the same time, strongly interdependent among themselves; thus, the modification of rules or explicit principles constantly creates new implicit principles; and finally, at least some of what we have called principles in the strict sense rest on what could be called the 'forms of life' of a society which, as everyone knows, are not easily identified in all their contours. But, in any case, to bring to light — reveal — the implicit principles of a legal order is, in our opinion, one of the central purposes of an ideological analysis of the law; the other one is to show the contradictory — or potentially contradictory — nature of the set of principles constituting a modern legal order.

2.3.3.2. With respect to the distinction between principles in the context of the primary system and principles in the context of the secondary system, what should probably be emphasized most from the perspective from which we are looking at norms right now is that, because of the special properties of principles, as opposed to rules, the former give to the law-applying organs a power (an ability to — positively or negatively — affect the interests of subjects) far greater than that given by rules. Therefore, the rising importance of principles in legal orders — as has often been observed — brings with it a progressive judicialization of the law.

3. *The explanatory, the justificatory and the legitimatory dimension of principles*

The last point to be treated here will be to show the most important functions principles fulfil in law. The starting point in this case will be the distinction between the dimensions of explanation, justification, legitimation and power-control present in principles. Each one of these functions — as we will presently see — is connected in a very special way with one of the three possible approaches to legal norms we have just distinguished: structural analysis especially underscores the function of explanation and systematization of the law;

regarding norms as reasons for action leads to the consideration of how they operate in justificatory legal reasoning; and, finally, looking at norms with respect to interests and power leads to the question about the use, the legitimation and the legal control of power.

3.1. Principles in legal explanations

One could think that asking to what extent principles enable us to explain what law in general, a legal order in particular, or a sector of such an order, is amounts to placing the question in the context of a legal science (with its different levels of abstraction). Principles would then be seen as the *parts* or instruments that allow us to account for a certain reality (the law, seen from different angles or levels of abstraction).

Now, principles have that explanatory function in at least two senses. First of all, because they can synthesize great quantities of information: reference to a few principles enables us to understand how a legal institution works, within the legal order as a whole and with respect to the social system. Principles — like scientific laws — are sentences facilitating a parsimonious description of some sector of reality (in this case, of the law); thus, they fulfil a didactic function, in a wide sense, of great importance. But secondly, and this is even more important, principles also enable us to understand the law — or different legal systems — not just as simple sets of standards, but as ordered sets, that is, sets that somehow make sense. Hence, if we know the principles of an institution or of some legal system we can even, to some extent, predict the solutions given to certain legal problems by specific provisions. This double capacity of principles, to present a part (or the whole) of a legal order in a succinct and ordered way, is precisely the capacity to what is usually called the systematization of the law. And since systematization is generally assumed to be the main function of legal science, principles obviously play a fundamental role here.

Although, of course, we do not want to deny this, we think it is important to stress that legal science — or, more precisely, what is usually called 'legal dogmatics' — is a normative discipline, and this not only in the obvious sense that its objects are norms, but also in two other senses (probably just as obvious, although this has not always been seen), namely: in the sense that its point of view is a normative one (the point of view of legal dogmatics is not an external point of view with respect to the norms) and in the sense that it has a function that could be called 'normative', since legal dogmatics does not restrict itself to the description of valid norms, but also proposes or suggests criteria for the solution of legal problems. In our view, in legal dogmatics, the systematization of a certain normative material is a central task, but not an end in itself; rather, it is a means for performing its most relevant social function: that of providing criteria for the application, interpretation and modification of the law. Actually, what characterizes legal dogmatics is not so much — or at least not only — its explanatory function, but rather its justificatory function.

3.2. Principles in legal reasoning

We will now turn to the justificatory dimension of legal principles, that is, in other words, to the role principles play in legal reasoning.

3.2.1. As a special kind of practical reasoning, legal reasoning is, of course, a complex activity that can entail elements of very different kinds. We will concentrate on its *normative* aspect. Here, one can say that the role of principles is different from that of rules because from some perspective, their contribution to reasoning seems to be rather modest, whereas seen from another point of view, one can say that principles are more important than rules.

Principles are *less* than rules in two senses. On the one hand, they do not have the advantage of rules, since they do not allow us to save time when we must decide a course of action. Because if a rule is accepted and it is applicable in a certain case, then one does not need to go through a process of weighing the reasons pro and con of a decision; a rule thus works like an element that reduces the complexity of the reasoning process. Principles, on the contrary, as we have seen, do not spare us the task of weighing. On the other hand, as premises to be used in practical arguments, principles have less force (are less conclusive) than rules. If someone accepts as a premise of an argument the rule "If X, then Y ought to be done", and also agrees that "X has occurred", then he must necessarily conclude that "Y ought to be done". But from the premise "E is an end to be reached to the highest possible degree" and the sentence that "D leads to E", we cannot reach the conclusion that "D ought to be done" (even if D is a behaviour that is not prohibited in the corresponding legal order), but only that "There is a reason for doing D"; the same applies to the premises "C is a valuable kind of behaviour" and "c is an action of type C" which do not allow us to reach a stronger conclusion than "There is a reason for doing c".

But principles are also *more* than rules, and again, in two senses. On the one hand, this is so because, since they are — or can be— stated in more general terms, they also come into play in a greater number of situations; that means that just as they have more explanatory power than rules, they also have greater justificatory scope. On the other hand, the lesser force of principles as premises of practical arguments is accompanied by a greater expansive force. Thus, for example, from the premises "All able-bodied men shall perform military service" and "If you have surgery, you will become able-bodied" we cannot conclude that "You ought to have surgery", and not even that "There is a reason for you to have surgery" (because it is perfectly possible that one does not wish to do military service, and therefore this does not count as a reason for oneself). But from the principle that "All Spaniards have a right to decent housing" together with the statement that "When loans for houses become cheaper, more people will be able to afford decent housing" it is perfectly possible to conclude at least that "There is a reason for the state to make loans for buying houses cheaper".

3.2.2. What has just been said about the different ways in which rules and principles work in legal reasoning seems to apply to principles in general, that is, without taking into account the triple classification of principles we have been using before. So now we should see whether these distinctions are relevant — and if so, in what ways — with respect to legal reasoning too.

3.2.2.1. First of all, it seems that there must be a difference between primary and secondary principles, to the extent that this distinction is understood to refer to whether principles are used by the general public or by judicial organs (when principles are used by lawyers, legal theorists, etc., this can, for the present purpose, be counted as judicial reasoning too). As has been said before, the difference is that for the former, principles generally are nothing but guides for action, whereas for judicial organs (as well as for lawyers and legal theorists) principles — like all other standards of the legal order — must serve not only to answer the question of what ought to be done, but also that of how to justify what has been done or what will be done, that is, legal decision-making.

Sometimes it is assumed that principles come into play in the justificatory reasoning of judicial organs only when they are confronted with a *hard case*, because in *easy cases* rules would be necessary and sufficient elements for justifying a decision. However, we believe that this way of looking at things is untenable, for the following reason. A case is easy precisely when the subsumption of certain facts under some rule is incontrovertible *in the light of the system of principles* giving sense to the institution or normative sector in question. And that they 'give sense' to it can, as has already been noted, itself mean two different things: they can be the values realization of which is assured by compliance with the rule, or the social objectives for whose attainment compliance with the rule is a means. But, whatever the meaning of 'giving sense' may be, it is always in the light of the explicit or implicit principles of the normative sector in question that we must determine whether a case is easy or hard.

This has important consequences because it contradicts a widely held view about how the law guides the conduct of judicial organs. That view can be summarized as follows: the law guides that conduct primarily and essentially through action rules, that is, through peremptory reasons (reasons — to say it once again — for issuing a decision whose content is based on the rule, and for precluding acceptance of any other reason as a ground for the content of the decision), and it requires deliberation by judicial organs only in marginal cases produced either by the deficits of ordinary language (cases falling in the twilight zone of the meaning of the terms used in the rule formulation) or by the relative indeterminacy of the legislator's purposes (cases of normative gaps and antinomies, as well as of judicial organs being given discretionary powers for resolving certain cases). On that conception, the main aspect of a judicial organ's law-guided behaviour is obedience to peremptory reasons; deliberation about non-peremptory reasons is peripherical or marginal.

Now, if a case can be regarded as easy — i. e., subsumible under some rule that must be accepted as a peremptory reason for resolving the case — only by taking into account principles, then the dimension of *obedience* to peremptory reasons can no longer appear as primary: *obedience* to such reasons requires prior deliberation and only takes place within the area thus delimited. The law, obviously, does not recognize any valid reason as a legitimate part of such deliberation: Except for some cases where it authorizes a judicial organ to invoke reasons other than those it contains itself, they must be reasons contained in the law itself, that is, explicit or implicit principles. Thus, we can speak of a *normatively guided deliberation* that also, if you wish, constitutes a kind of 'obedience', but — and this is the important part — which differs substantially from the obedience consisting in following a peremptory reason.

All this can probably be summarized by saying that for judicial organs the law constitutes an *exclusionary system* on two levels and in two senses. In a first sense — and on a first level —, insofar as it imposes on the judicial organs the duty to find a balance of reasons, where only legal standards count as reasons and where other reasons may be taken into account only to the extent that the legal standards themselves allow this. In a second sense — and on a second level —, insofar as that balance of reasons in most, though not in all, cases requires the use of a legal rule, that is, a peremptory reason, as the basis for decision. Thus, we can divide cases into two groups: those the solution of which is based on the balance of legal reasons that enter into a judicial organ's deliberation; and those where that balance of reasons requires to suspend deliberation and to adopt a peremptory reason as the basis of decision.[5]

3.2.2.2. As for the distinction between the argumentative use of principles in the strict sense and policies, the fundamental things have already been said above, when we contrasted rightness reasons with reasons of a utilitarian kind. We can now add that we agree with Summers' assertion (1978) that *rightness reasons* and *goal reasons* are the main types of substantive reasons (which he distinguishes from authoritative reasons) which, in turn, are the central element of legal reasoning in hard cases.

3.2.2.3. Finally, with respect to the distinction between explicit and implicit principles, there can be no doubt that the latter imply a higher complexity than the former because they not only give rise to the question of how they should be used as premises of a legal argument, but also of how one can justify their adoption as such premises. In that sense, in our view, to say that "X is an implicit principle" of some institution, normative sector or legal system is the same as saying that the corresponding rules and explicit principles are consistent with X and that, when considered as linguistic formulations, they should be understood as having a propositional content that also is consistent with X. But, as is well-known, the requirement of consistency with the normative material often can be satisfied by principle X as well as by other principles. In that sense, the problem of identifying implicit legal principles coincides with the

---

[5] In the first case, the deliberation of the judicial organ leads to the formulation of a rule. This means that in a hard case, once it has been accepted that the case must be solved by a balancing of reasons, the judge proceeds in two steps: the first is to go from the principles to a rule expressing, for the class of situations in question, the result of that balancing or weighing of principles; the second is that, since now a rule is already there, this rule — that is, a peremptory reason — is adopted as the basis for decision. The difference with easy cases is that here the rule (the peremptory reason) has been formulated — on the basis of a balancing of non-peremptory legal reasons — by the judicial organ itself. We should add that with this we do not claim to have given a description of the psychological process going on within the judicial organ but only a conceptual reconstruction.

general problem of legal interpretation. And that is a problem which clearly exceeds the limits of our present work.

3.3. Principles, control and legitimation of power

The question we must confront now — that of how legal principles affect the use, legitimation and control of power — is one of the aspects of the more general question about the social functions of law. Obviously, principles are connected with the realization of all social functions performed by the law; but here we are interested only in the one concerning the use of power for the attainment of certain social objectives (for shaping the conditions of life in society) and its legitimation and control; in a way, these are different sides of one and the same reality.

The first one of these functions — or, if you prefer, the positive side of this complex function — has to do with the new reality of our modern state — the welfare state — who creates a law that is no longer restricted to the classical functions of coercion and protection, but also claims to operate as a technology of social engineering. This is precisely what is reflected in the rising importance given to policies in all legal orders of developed societies.

Now, together with this tendency, so to speak, to increase the power of the law in society, there is also an increasing demand that this power be restricted by moral criteria usually incorporated into legal orders in the form of human or basic rights. Therefore, legal principles — in this case, especially legal principles in the strict sense — are more and more explicitly formulated in the declarations of rights to be found in modern constitutions. Of course, the need to legitimize and control power is not new. But what is perhaps new is how this need is treated. As Aulis Aarnio observed, "in Western legal cultures, the belief in authorities has suffered greatly over the past decades" (Aarnio 1987, XV). In turn, this has led to stronger demands concerning the justification of their decisions. Today, the main prerequisite for the exercise of public powers to be considered legitimate is that they must be able to pursue — and attain — social objectives without violating the basic rights of individuals. Attainment of this difficult balance — which has become the regulative principle of contemporary jurists — will depend to a great part on the development of an adequate theory and practice of legal principles.

APPENDIX TO CHAPTER I
REPLY TO OUR CRITICS

Previous versions of our paper on legal principles have been subjected to a number of criticisms, most of them expressed orally in several seminars where we had the chance to discuss our ideas.[1]

At one of these seminars, the *II Spanish-Finnish Seminar on Legal Theory* held in Tampere in September 1992, Aleksander Peczenik presented a paper (Peczenik 1992) that turned out to be a global critique of our conception of legal principles. This is also the case with the first chapter of Luis Prieto Sanchís' book on principles in law (Prieto 1992),[2] although, of course, his discussion is not limited to the critique of our theses, and includes all the different conceptions of legal principles to be found in contemporary legal philosophy.

We will try to reply to both critics. It should be noted at the beginning that both Peczenik and Prieto criticize the central pillars of our contribution, and that in their texts — with obvious differences of emphasis — we can also find the basic tenets of the critiques voiced in oral discussions by other legal philosophers. Thus, trying to answer to their objections will be an excellent testing ground for us.

As we understand it, the core of Prieto's and Peczenik's critiques can be divided into the following two points: *a)* one refers to our understanding of mandatory legal rules (action rules or end rules) as peremptory reasons and to our idea that such rules differ from principles in the strict sense in that the former present cases in 'closed' form while the latter present them in 'open' form; *b)* the other refers to our conception of principles in the strict sense as norms requiring full compliance, that is, norms that do not allow compliance by degree. One could think that *a)* contains at least two different questions, but we believe that in our presentation as well as in the critiques by Prieto and Peczenik we are dealing rather with two different aspects of one and the same

---

[1] That paper has now become the first chapter of the present book. A first version of it was published under the title 'On principles and rules' in *Doxa* 10, and discussed in a seminar on current problems of legal theory in the United States and Spain, held at the *Centro de Estudios Constitucionales* in Madrid in June 1991. An expanded later version under the title 'Three Approaches to Legal Principles' was presented in 1992 at the II Spanish-Finnish Seminar on Legal Theory and published in Italian ('Tre approcci ai principi di diritto') in the 1993 volume of *Analisi e diritto*.
[2] A first version of our reply to Prieto was published in *Doxa* 12; to this, Prieto has, in turn, replied in the following issue of the same journal (cf. Prieto 1993).

question, because the fact that rules can function as peremptory reasons implies, as a necessary condition, that they determine their conditions of application in 'closed' form.

1. *Mandatory rules as peremptory reasons and principles as non-peremptory reasons; the 'closed' or 'open' configuration of the conditions of application*

In chapter I we have adoptedAlchourrón/Bulygin's characterization of norms as correlations between generic cases (sets of properties) and normative solutions, on the one hand, and Hart's characterization of authoritative legal reasons as peremptory and content-independent reasons, on the other. In doing this, we explicitly limited the scope of these characterizations to *mandatory legal rules*, that is, we left out other types of rules (like permissive or power-conferring rules) which we did not wish to consider there, and we also left out principles, which — though they were clearly part of our topic — we tried to characterize *structurally*, following the model of Alchourrón/Bulygin, and from the perspective we called *functional*, following Hart's characterization. Our critics have very well understood the link between the two questions: in fact, if rules can be not only first-order reasons to perform a required action, but also second-order reasons — in Hart's words — "to preclude or cut off any independent deliberation by the hearer of the merits pro and con of doing the act", this is so because they articulate their conditions of application in 'closed' form. Starting from different perspectives, Prieto and Peczenik pursue the same line of argument: that the alleged 'closed' configuration of the conditions of applications is not really closed, and therefore, rules cannot be peremptory reasons.

1.1. Prieto's critique

Let us begin with Prieto's critique which we will quote *in extenso* and then analyse step by step. According to a certain interpretation, Prieto writes,

"what distinguishes rules from principles is said to be that *a priori*, before their application, we can say precisely in what cases a norm should be observed, since the order foresees, or should foresee, all possible exceptions to its application, whereas principles do not and cannot have such a clause [...] In consequence, the difference is said to be that while we can know with certainty when the solution foreseen in a norm will be imposed, there is a margin of doubt concerning the solution supported by a principle."

Among those who hold this thesis, Prieto names Dworkin and ourselves: "Although they differ from Dworkin in important ways, I think that among us M. Atienza and J. Ruiz Manero follow a similar line." He quotes our thesis that rules give their conditions of application in 'closed' form while principles do this in 'open' form, and adds that "on this criterion of structural distinction, Atienza and Ruiz Manero superimpose another, functional one", i. e., the consideration that, while rules constitute peremptory reasons, principles are merely first-order reasons. "If I do not misunderstand them", he goes on, "this means that rules must contain all possible exceptions to their application, and therefore once a norm has been selected there is no need for weighing any other norm or principle; principles, in contrast, are characterized by competing against other principles." According to Prieto, this way of understanding the distinction between rules and principles is fundamentally flawed because it is based on "a false presentation of rules":

"[A]ctually, in legal reasoning rules often appear as hermeneutic criteria and not as specific rules exhaustively treating the case at hand; they also must often be combined with other legal standards (principles or norms), thus changing their extent and their very scope of application; and finally, their meaning is always open-textured or potentially vague, so they do not completely and absolutely enumerate a catalogue of possible exceptions for their application. Thus, the indicated position seems to be overly confident in the logical finiteness, if not of the entire system, at least of the system of rules. [Because] it seems rather improbable that someone could be capable of completely enumerating all possible exceptions to the application of a normative solution: on the one hand, because through legislation or jurisprudence new exceptions can always arise — e. g., through the appearance of new exempting or extenuating circumstances — and, on the other, because changes in the interpretation of the law also can, in the guise of old exceptions, open the door to new circumstances — e. g., if it is considered that parliamentary immunity excluding or conditioning the exercise of criminal action extends to the members of a regional legislative body. And what's more, this is not only a matter of gaps, or of not knowing all exceptions, but also one of excess, that is, of not knowing a priori what cases not mentioned in the norm nevertheless ought to receive the same treatment, by the method of analogy. In fact, the question of when we should not apply a norm because there is an exception as well as that of how far we should go in its application, by way of analogy, seems to be much more complex than the ideas of 'all or nothing' or of 'peremptory reasons' imply [...] And, paradoxically, as Alexy has observed, the existence of principles explains to a great extent why the 'all or nothing' view fits neither rules nor principles. Because if one asserts that the exceptions to principles, and therefore also their conditions of application, cannot be enumerated, and if one asserts, in contrast, that principles can ground exceptions from a rule, then the logical consequence is that we also cannot know the cases of exception from a rule."

In order to reply to this critique, we must try to separate the different questions implied in it:

1. A first distinction that is important here, and which Prieto seems to pass over, is the one concerning the difference between generic and individual cases. Generic cases are nothing but sets of properties. Therefore, a *system of rules* that contains an exclusive general rule permitting everything that does not fall under the prohibitions particular rules correlate with the descriptions of the generic cases they themselves contain, would have no *normative gaps*, i. e., would be a *complete* — or, as Prieto prefers to say, a *logically finite* — *system* with respect to the generic cases. Quite another matter is that in the description of generic cases there may be problems of semantic indeterminacy such that with respect to an individual case (or a case less generic than that described in the rule) doubts may arise about what generic case it belongs to. Although such a system would have no *normative gaps*, there would be instances of what Alchourrón and Bulygin have called *gaps of recognition* (that is, the kind of doubts that arise because of the problems of semantic indeterminacy just referred to). In 'On principles and rules' — and now in ch. I of this book — we have been very much aware of this distinction between generic cases and individual cases, and we have pointed out that we adopted "the conception of norms to be found in Alchourrón and Bulygin's *Normative Systems* (1971), where norms are seen as sentences correlating generic cases (sets of properties) to solutions (that is, to the normative qualification of some conduct)" and also that the "semantic indeterminacies" legal provisions may present "do not affect the character of a rule". Thus, Prieto's totally plausible assertion that "their meaning [of rules] is always open-textured or potentially vague" does not affect the question of the 'closed' configuration of their conditions of application. This is so because what, in our view, rules do present in 'closed' form are the generic cases that constitute their (generic) conditions of application, whereas the question of the open texture or potential vagueness arises in another context, namely, that of the subsumption of individual cases under such generic descriptions. And this is precisely the context of the example given by Prieto when he says that "changes in the interpretation of the law also can, in the guise of old exceptions, open the door to new circumstances — e. g., if it is considered that parliamentary immunity [...] extends to the members of a regional legislative body". In that case, we are dealing with an instance of vagueness of the word 'parliamentarian' that gives rise to doubts — in the end, decided positively —

about whether or not the members of a regional legislative body fall under the description of the generic case ("being a parliamentarian") of the rule granting immunity to those who have that property. But this has nothing to do with whether or not the corresponding rule configures its (generic) conditions of application in 'closed' form — which, of course, it does: if someone is a parliamentarian, he or she has that immunity; if not, then he or she does not have it.

The idea that the conditions of application in rules are configured in a closed way (in contrast to what is the case with principles) can be elucidated, somewhat paradoxically, by looking at the work of authors who, like Alchourrón and Bulygin, do not have much sympathy for the rules-principles distinction. In an article first published in 1988 ('Conditionality and the representation of legal norms', in Alchourrón/Bulygin 1991), Alchourrón asks how legal norms can be adequately (that is, in a way compatible with the common intuitive interpretation) formalized. The problem is that norms make the stipulation of obligations or the conferring of rights depend on the satisfaction of certain conditions. However, norms usually do not explicitly mention all the conditions required for the corresponding obligation or right to arise; generally, they tend to omit those negative conditions whose presence prevents a right or obligation from emerging. This leads to problems of the following kind. Suppose a normative system includes the two norms:

a) Judges shall punish those who have committed murder.
b) Judges shall not punish persons under age.

Using bivalent propositional logic and standard deontic logic, they would be expressed through the following logical formula:

c) $p \rightarrow Oq$.
d) $r \rightarrow O\text{-}q$.

But from this it follows that in the case when the conditions of application of the first norm (having committed murder: p) and the second norm (being under age: r) both hold, then judges face a conflict of obligations. Because (with the law of strengthening the antecedent) from *c)* one can derive $p.r \rightarrow Oq$, and from *d)* one can derive $p.r \rightarrow O\text{-}q$. That means that the legal system forces a judge to conclude that he has the obligation to punish as well as the obligation not to punish those who commit murder while being under age.

In order to solve this problem, Alchourrón, on the one hand, distinguishes between the language of the norms of the system and the meta-language that describes the obligations resulting from the system in question, and, on the other, he introduces a relation that under certain conditions gives priority and preference to one norm over another. In his view, then, there is in fact a conflict of obligations — on the level of norms — in the case just described, but the conflict is solved — at the level of the meta-language — because norm *d)* has priority over norm *c)*. Thus, the answer to the question "What should a judge do in the case of *p.r*?" is simply that he should not punish.

Now, Alchourrón shows that in legal systems which have the type of ordering relation just presented (a transitive and asymmetric relation), under no condition (except an impossible one) can there be conflicts of obligation, since what is obligatory is determined not only by the conceptual content of the norms but also by their rank order. What he does not seem to see, however, is that in a legal system there may be norms stipulating incompatible normative solutions that do not stand in such a relation, and where this is not a simple case of antinomy either. And that is precisely what, in our view, happens with principles: they stipulate obligations, configuring their conditions of application in what we have called an *open* way — '*p* is obligatory unless that obligation is overruled by a principle which in the case at hand has higher weight' — and *without establishing an order*, because the system does not predetermine a rank order (the 'relative weight') for the case of conflicting principles. Therefore, in order to solve a case that involves principles, an intermediate operation is required, that is, a new rule must be stipulated (on the basis of those principles). This operation, which consists in transforming principles into rules, is usually called *specification*. We will come back to it later, in the context of what we have called the *expansive force* of principles which, as Peczenik has pointed out to us, we did not sufficiently explain.

2. The assertion that rules configure their conditions of application in 'open' form, thus, does not deny, but rather is perfectly compatible with admitting that the description of those conditions may present a zone of penumbra where the subsumption of particular individual cases may be doubtful. And, of course, we also do not think that the fact that "through legislation or jurisprudence there can always arise new exceptions" to the applicability of a rule poses any problem for our conception. That the normative authorities of a system may change that system by restricting the conditions of application of some normative solution is obvious; the idea that rules present their conditions of application in

'closed' form is, of course, limited to the rules existing at a given moment and does not imply any thesis about changes in the normative system.

3. Although it is somewhat more of a problem, it also does not seem to bring any special difficulty for our conception if it is admitted that a case that is not normatively regulated by the system is solved by analogy. What the admissibility of the argument by analogy shows is precisely that when an individual case cannot be subsumed under the conditions of application of an existing rule (and precisely because such conditions of application are of a 'closed' character), as a basis for his decision, the judge must construct a general rule correlating other conditions of application with the same normative solution which an existing rule correlates with conditions of application the judge sees as substantially similar. And such a relation of similarity can not be asserted (nor denied) on any basis other than the principle that explains and justifies the existing rule. That means that the argument by analogy always implies the use of principles, although it does not mean that recourse to analogy and to principles are the same thing. Actually, analogy is an argument — or, more precisely, a structure of reasoning —, whereas principles are a material that must necessarily be used in that type of reasoning.[3]

4. Prieto does not object against our thesis that principles present their conditions of application in 'open' form. But he asserts that this has a consequence which invalidates our characterization of rules, when he writes: "[...] if one asserts that the exceptions to principles, and therefore also their conditions of application, cannot be enumerated, and if one asserts, in contrast, that principles can ground exceptions from a rule, then the logical consequence is that we also cannot know the cases of exception from a rule". Here, Prieto is right; but that does not have the fatal consequences for our proposal that he claims, although it does, of course, force us to specify — as we have already done in ch. I — the scope of our thesis that rules present their conditions of application in 'closed' form and constitute — when those conditions of application apply — peremptory reasons.

In our view, it is true that the applicability of any rule is conditioned on its application not being in conflict with a principle that has greater weight with respect to the relevant properties of the case. Once this is recognized, we could, of course, conclude — as realists do — that rules are merely cute little toys and

---

[3] Cf. on this point Atienza 1986.

that legal reasoning is always radically open, and then dedicate ourselves to verbal games about indeterminacy, like those the authors of *Critical Legal Studies* are so fond of. The problem with this kind of orientation, as has often been explained, is that it clashes with the evidence that in the vast majority of individual cases taken to the courts, the subsumption under the generic case treated in some general rule — and the issuing of a judicial decision thus grounded — does not give rise to any kind of controversy in the legal community.

In order to be able to account for both circumstances — the subordination of the applicability of rules to principles and the fact that the great majority of cases is solved through an uncontroversial application of rules — we proposed to understand the way in which the law guides the reasoning of its organs of application as a two-level structure: on a first level, we said, the law "imposes on the judicial organs the duty to find a balance of reasons, where only legal standards count as reasons and where other reasons may be taken into account only to the extent that the legal standards themselves allow this"; on a second level, "that balance of reasons in most, though not in all, cases requires the use of a legal rule, that is, a peremptory reason, as the basis for decision". Thus, that rules present their conditions of application in 'closed' form (and that rules themselves operate as peremptory reasons), in our view, applies only at the second level.[4] Of course, this second level is reached when the principle

---

[4] We believe that a similar position (although expressed in a different terminology) can be found in the three theories of rules that are probably most influential today: those of Raz, Regan and Schauer.

Thus, we do not believe that Raz means something very different when he writes that "[r]ules are, metaphorically speaking, expressions of compromises, of judgements about the outcome of conflicts. Here talk of exceptions comes into its own. Characteristically, cases are 'simply' outside the scope of the rule if the main reasons which support the rule do not apply to such cases. Cases fall under an exception to the rule when some of the main reasons for the rule apply to them, but the 'compromise reflected in the rule' deems other, conflicting reasons to prevail" (Raz 1990, 187).

Similarly, Donald Regan observes that rules must be regarded as neither absolutely transparent nor absolutely opaque. A rule would be treated as 'absolutely transparent' if one would think that one should do what the rule commands only when one is completely sure. after weighing all the reasons that apply in the case in question, that that is in fact the correct action. A rule would be treated as 'absolutely opaque' if one would think that one should always, whatever the reasons applying to the case at hand, do what the rule commands (Regan 1989, 1004-1013). As Juan Carlos Bayón has remarked, it is clear that treating a rule as absolutely transparent would make it superfluous as an instrument for decision-making, whereas treating it as absolutely opaque would certainly be irrational (Bayón 1991b, 51 f.).

Finally, another thesis in the same direction is sustained by Frederick Schauer. He points out that a central feature of rules is the circumstance that they are potentially under-inclusive or over-inclusive with respect to the underlying reasons, that is, there may be cases that are not included, though they should be, and others which, in turn, should not be, but are included (Schauer 1991).

that 'One ought to do what legal rules prescribe' is not overruled, on the first level, by a principle that in the case at hand has higher weight. But when that principle is not overruled, then in the reasoning of the judicial organ the legal rule under which the individual case is subsumed functions as a peremptory reason for issuing a decision whose content is in accordance with the rule.

1.2. Peczenik's critique

Aleksander Peczenik rejects our idea that principles are merely first-order reasons for deciding in a certain way, whereas rules are 'peremptory' or 'protected' reasons. Thus, he writes that "not only principles, but also some rules create merely *prima facie* duties and, therefore, require deliberation". And he goes on:

"I think that 'legal standards' allow us to take into account all morally relevant considerations [...] Established legal norms constitute *prima facie* reasons that must be weighed and balanced against other reasons. Those *prima facie* reasons are first-order reasons for performing a certain action, *A*, and, at the same time, second-order reasons. In this latter capacity, they indicate that the reasons for not doing *A* may prevail only if they are particularly strong, that is, clearly stronger than they would need to be in a free moral debate."

These theses of Peczenik — which in the paper to which we are replying here only are stated — are developed more fully in his important book *On Law and Reason* (Peczenik 1989) to which he himself refers. In that book, in a section titled 'Weighing Rules', one can read the following:

"However, not only principles but also some *rules* create a merely *prima facie* duty. This is true about both moral and legal rules. For example, one ought not to kill people. The moral rule forbids *prima facie* all killing but to state that a given individual, all things considered, ought not to be killed, one must also pay attention to other rules, stipulating exceptions; for instance, in a defensive war, one may kill the aggressors. [...] *All* socially established legal norms, expressed in statutes, precedents etc., have a merely *prima facie* character. The step from *prima-facie* legal rules to the all-things-considered legal (and moral) obligations, claims etc. involves evaluative interpretation, that is, weighing and balancing." (Peczenik 1989, 80 f.)

In our view, several objections can be raised against this conception of Peczenik:

1. The first is that his example of a *rule* creating merely a *prima facie* duty (the prohibition to kill), according to our definition, is not a rule, but a *principle*. This is so because that norm does not determine its conditions of application

other than in what we have called *open* form (which we regard as characteristic for principles): "It is prohibited to kill, unless there is another principle pertinent to the case at hand that has more weight". Of course, one can object against drawing a conceptual distinction between types of legal norms, like the one we have drawn between principles and rules, on the basis of whether the conditions of application are determined in an 'open' or 'closed' way, by saying, e. g., that this characteristic is not very relevant, that it does not reflect ordinary usage of the terms 'rule' and 'principle', etc. But if one accepts our distinction, then it is clear that (in contrast to the *rules* on homicide contained in the Criminal Code) 'Thou shalt not kill' is not a rule, but a principle. Besides, even according to Peczenik's own characterization of the distinction between rules and principles one would have to conclude that the prohibition to kill is to be understood as a principle. This is what seems to be implied by his remarks that "Unlike a principle, the rule [...] does not express a single value but a compromise of many values (and corresponding principles)." (Peczenik 1989, 81) and that "the main source of the justificatory force of principles is their one-to-one link to the corresponding values. Every principle corresponds to some value [...]" (Peczenik 1992).

2. The second objection refers to Peczenik's thesis that legal standards allow one to take into account all morally relevant considerations, that established legal norms constitute *prima facie* reasons which in their dimension of second-order reasons only indicate that the reasons for not performing the action required by them can prevail only if they are especially strong. In our view, this thesis is ambiguous, because it can be understood *i)* as a thesis that intends to account for *the claims of the law*, or *ii)* as a thesis that intends to account for *the attitude a practically rational subject should adopt with respect to such claims*. Neither of the two cases seems acceptable to us. If it is meant to refer to the claims of the law, it is clearly false. To give an example that can hardly be controversial: To restrict someone's basic rights on the grounds of his race contradicts the morally relevant consideration that all human beings should be granted such basic rights for the simple fact that they are human beings. But, obviously, there have been legal orders that did not allow one to take this consideration into account. And, on the other hand, if Peczenik's thesis is understood as referring to the attitude a practically rational subject should take with respect to the claims of the law, then the thesis concedes too much. To say that a practically rational subject should always consider existing legal norms as

prima facie reasons for performing the actions prescribed by them would mean that a practically rational subject should accept that there is a *prima facie* general moral obligation to obey the law. This is something Aleksander Peczenik has defended repeatedly (cf. 1989, 238 ff.; 1990, 96 ff.), but in our view, his arguments are not convincing. As he himself says, the "central point" of his theory is the following: "There exists a general *prima facie* moral obligation to obey the law because general disobedience would create chaos" (1989, 246). Peczenik justifies this thesis not by an appeal to the causal effect acts of disobedience would have on the maintenance of the system, but by an appeal to the requirement of universalization. Peczenik admits that there are acts of disobedience that do not at all increase the probability of other acts of disobedience, and he thinks that Raz's critique of attempts to ground the thesis of the existence of a *prima facie* general moral obligation to obey the law on the causal effect (of a 'bad example') is correct. Peczenik himself grounds his thesis on the following "universal premise" that "is a consequence of [the] universalisable character of morality": "I have a *prima facie* moral obligation to act in such a way that my action could be repeated by everybody without creating morally wrong consequences" (Peczenik 1989, 246). But if the principle according to which I act is that 'It is morally legitimate to disobey the law when my act of disobedience does not in any way increase the chance of other acts of disobedience and, therefore, does not causally affect the conservation of the system', then it is analytically true that everyone could act on that principle (that is, that the principle could be universalized) without morally incorrect consequences (i. e., an increase in the probability of other acts of disobedience, and, in consequence, the collapse of the system) (cf. on this point Bayón 1991a, 708 f.). In fact, then, Peczenik's argument does not serve as a justification of his thesis, just as it would not serve as a justification of the judgment that 'There is a *prima facie* general obligation not to drive by kilometer X of road Z at H hours' to say that if everyone would drive by kilometer X of road Z at H hours, traffic would collapse. But this already surpasses the limits of our present topic concerning principles in the law.

2. Principles and full compliance

Though their positions differ, Prieto and Peczenik agree in rejecting our thesis that principles in the strict sense, as opposed to policies, do not permit compliance in degrees.

2.1. Prieto's position

Under the heading 'Normative characterization of principles', Prieto opposes the idea — based on some suggestions by Dworkin — that principles are never sufficient reasons for decision, that they "can guide an uncertain normative interpretation, but can *never* by themselves offer the solution of a case" since "they neither offer nor refrain from offering a categorical answer but rather 'control' (by expanding or restricting them) the solutions deduced from the entire set of norms".[5] "I think", Prieto writes, "that this is an acceptable interpretation of Dworkin's words, but that it is also probably a wrong opinion." In order to argue why this opinion is wrong, Prieto invokes the jurisprudence of the Spanish Constitutional Court as well as Dworkin's own examples. As for the first, he reminds us that the Constitutional Court has declared that, whenever there is an irreducible opposition between statutes and (constitutional) principles, the principles share the derogative force of the Constitution, which means that "at least in certain cases a principle represents the only foundation of a decision", this being so when the Constitutional Court declares a legislative provision unconstitutional because it violates a constitutional principle. And as for Dworkin's famous example which Carrió has called "the case of the grandson in a hurry", Prieto writes that "if, according to New York law, descendants inherit from their elders and no rule excludes the case of the grandson who murders his grandfather, then this means that if a court denies that consequence in the name of the principle *nemine dolus suus prodesse debet*, that principle is the only foundation of its decision; one cannot say that the principle has been taken into account in order to tip the balance in favor of one or the other normative solution — both of them supported by rules — since there are only two solutions here: the (only) one that should be adopted if the rules are respected, and the other one imposed by principles."

Up to this point, we totally agree with Prieto. We think that what his two examples show is that once the prevalence of some principle in a certain case has been determined, the principle in question requires full compliance, which in the examples mentioned implies the annulment of the unconstitutional legislative provision and the rejection of the grandson's claim to take possession of his grandfather's inheritance. However, a few pages later, Prieto adheres to what could be called a weakened version of Alexy's conception of a 'mandate

---

[5] Among those who have sustained such a conception of principles, Prieto names his own work (Prieto 1985) and that of J. Ruiz Manero (1990).

of optimization *[Optimierungsgebot]*', by defending not only the idea of compliance by degrees but also that principles merely require a 'reasonable' degree of compliance:

"[T]he idea of the mandate of optimization [...] also does not seem to serve the purpose of distinguishing principles in the strict sense from policies, as Atienza and Ruiz Manero sustain, because when there is a conflict principles too can be optimized; maybe one could say that policies *naturally* lead to mandates of optimization, but that does not imply that this technique is totally excluded from principles in the strict sense. I will try to explain this with an example. Imagine a clause in a will where a father imposes as a condition for his son to be given possession of his inheritance (the legal and the free portion) that he must divorce his wife who is Jewish. To solve the case, the judge must keep in mind two principles: that of equality, prohibiting discrimination on religious or racial grounds; and that of autonomy of the will which protects the intentions of the testator. Now, it would not be strange at all if in this case the idea of optimization were simply discarded and full preference were given either to the principle of equality or — which I think would be the worse solution — to that of autonomy of the will; but it would also not be surprising if a judge were to balance the two principles and say that the will is null with respect to the legal portion, and valid with respect to the portion the testator could freely dispose of.
[...] It is undoubtedly true that gradual compliance with some standards and judgments of optimization are ideas that can be found in legal reasoning, but it may be premature to say that principles always command that something be done to the highest possible degree. As the Constitutional Court has declared, 'the Constitution is a frame of reference sufficiently broad for political options of very diverse kinds to be able to fit under it', and without any doubt the programs of those options can contain different ideas about the degree to which constitutional principles or policies should be achieved. Therefore, from the perspective of the judge, we should speak not so much of a judgment of optimization as of a judgment of reasonableness which, accepting the idea of compliance in degrees, indicates the level of compliance or satisfaction below which some norm or policy becomes intolerable."

Prieto himself has developed this last idea further in his article 'Notes on constitutional interpretation' (1991). There, under the heading of 'Some peculiarities of constitutional justice', he writes:

"The difference between legal and constitutional interpretation lies not only in the peculiarities of its object, but also in the function generally attributed to the organs in charge of it. With respect to this, one of the characteristics of the ordinary judge is what could be called the 'unity of a just solution', that is, the institutional requirement that each particular case only have one right solution, whereas the task of constitutional justice is to indicate what interpretations cannot be tolerated, rather than to stipulate the 'best' or 'only' possible answer. [...] This different function [...] also contributes to understanding the responsibility of the decision in another way. Thus, although it is, of course, a fiction, the ordinary judge can 'blame' the sense of the decision on the legislator [...] In contrast, the constitutional interpreter does not really look for a solution to the case, but for the contours of an area of legal permission within which other legal operators will adopt a solution in accordance with political (legislator) or legal (judge) criteria; thus, his way of reasoning must not be adjusted to the canons of subsumption, but to those of reasonableness [...] In other words, the kind of reasoning of an ordinary judge implies that the decision is conceived 'as if' it came from the legislator, whereas the pattern of reasoning of a constitutional judge who must define the — more or less wide — range of what is permitted requires that the interpreter take on greater responsibility for the decision. In that sense, I think A. Carrasco is right in saying that the characteristic method of the constitutional interpreter is halfways between what he calls the strict deduction typical of ordinary justice and a political judgment of optimization. On the one hand, because of the very nature of its activities, the Constitutional Court cannot merely verify an act of subsumption, since in most cases that which must be judged does not correspond to facts stipulated in any law, nor can the parameters for the judgment refrain from jointly balancing and weighing principles and rules; and, on the other, it must also exercise 'self-restraint' in order not to hand down a judgment of optimization that would imply a decision on which is the 'best' interpretation of the constitutional text and, therefore, eliminate Parliament's margin of evaluation." (Prieto 1991, 176 ff.)

In our view, what is wrong with all those considerations about understanding constitutional jurisdiction is something that has to do with the — in our opinion, mistaken — way in which Prieto understands the role of principles in the reasoning of judicial organs. Stated briefly, the mistake is basically that Prieto presents as peculiar characteristics of constitutional justice what, in fact, are characteristics of any exercise of the judicial function where what is judged are acts constituting an exercise of normative powers granting the powerholder a margin of discretion. Let us look at this somewhat more closely. Prieto says that "the task of constitutional justice is to indicate what interpretations cannot be tolerated, rather than to stipulate the 'best' or 'only' possible answer. [...] The constitutional interpreter does not really look for a solution to the case, but for the contours of an area of legal permission [...]; thus, his way of reasoning must not be adjusted to the canons of subsumption, but to those of reasonableness". Of course, it is not the task of constitutional justice to spell out what is, from the point of view of the Constitution, the best possible statute; but it also is not the task of a judge at a court for administrative law to spell out what is the best possible regulation, from the perspective of the statute such a regulation develops, nor is it the task of a judge for lawsuits under civil law to determine which is the best possible contract or will from the perspective of the Civil Code. Such things simply do not exist. It is neither possible to derive the 'best' legislation or the 'only' admissible legislation on some matter from the Constitution, nor to derive from the Civil Code the 'best', or the 'only' admissible, contract or will. What the Constitution and the Civil Code do is, respectively, to confer the public normative power to legislate and the private normative powers to make contracts and to lay down a will, while at the same time imposing certain restrictions on the exercise of those normative powers. And what the Constitutional Court does when it judges the constitutionality of a law, or what a judge at a civil court does when he judges the validity of a contract or a will, is to judge whether or not the holders of the corresponding normative powers have used them in a way that violates those restrictions. And in this respect, the Constitutional Court as well as the civil judge must — in contrast to what Prieto seems to think about the first case — look for a solution of the case: If the legislator has used his public normative power, violating the restrictions imposed by the Constitution, then the Constitutional Court must declare the law unconstitutional in all aspects affected by that violation; and if the testator has used his private normative power, violating the restrictions imposed by the Civil Code, the judge also must declare the will invalid in all aspects affected by that violation. On the other hand, if either one of them has exercised the respective normative powers in a way that does not violate those restrictions, the Constitutional Court or the judge must declare the statute constitutional or the will valid, even if the provisions contained in them may seem extravagant. Just as the Constitutional Court — as Prieto writes — must "exercise 'self-restraint' in order not to hand down a judgment of optimization" about what the best possible statute would be, the civil judge too must exercise 'self-restraint' in order not to issue a judgment of optimization about what would be the best possible or the most reasonable will, or something of that kind. Both — the Constitutional Court as well as the civil judge — must reason for their resolution in the same way: by showing that the statute (or the will) has violated (or has not violated) the restrictions imposed by the Constitution (or the Civil Code). In this context, that the Constitutional Court is referred to the Constitution and the ordinary judge to the Civil Code is not a relevant difference. And the first as well as the second can be facing a case where they must take into account rules as well as principles, as Prieto's own example about the will subjected to the condition of divorce from the Jewish wife shows. This example deserves some more attention.

According to Prieto's reconstruction, the judge in that case could decide on one of three solutions: to regard the will as null, which would mean to give "full preference to the principle of equality"; to regard the will as entirely valid, which would mean to give full preference to the principle of autonomy of will; or to regard the will as null with respect to the legal part, and valid with respect to the part of free disposition, which would mean that the judge tries to "balance the two principles." In our view, this reconstruction is wrong. Rather, as we see it, things are as follows: The principle of autonomy of will and the principle of equality come into conflict only with respect to the part of free disposition. With respect to the legal part, the case is covered by the rule that descendants can only be deprived of their part for certain reasons, to which being married to a person of the Jewish faith does not belong. That rule can be seen as an expression of the prevalence — laid down by the legislator with respect to part of an inheritance — of the principle of protection of the interests of descendants over the principle of respect for the autonomy of the will of the testator. So, unless one argues (as Prieto does not) that the rank order of principles expressed in the rule about the legal part of an inheritance clashes, in the case at hand, with some other principle, with respect to the legal part the case is an easy case, because the application of a rule whose generic conditions of application completely cover the specific circumstances of the case can hardly be objected. Things are different, though, with respect to the part of free disposition. And this is so because, in contrast to the legal part, here there is no rule that would determine the prevalence between the principle of autonomy of will and some other possibly conflicting principle (in our case, the principle of equality). This, precisely, is the reason why the judge must balance the two principles, that is, construct a rule that stipulates such prevalence. But here, we have two and only two possibilities: either the principle of autonomy of the will prevails (and the corresponding clauses in the will therefore are valid), or the principle of equality prevails (and those clauses are null). In other words: once the prevalence of one of the two principles is established, full compliance is required, that is, the corresponding clauses in the will must be declared either valid or null. *Tertium non datur.*

2.2. Peczenik's position

Peczenik offers two arguments in defence of Alexy's conception of principles as mandates of optimization and against our thesis that principles in the strict sense require full compliance. The first one is that the example of a principle in the strict sense which we used to illustrate our thesis (art. 14 of the Spanish Constitution) is not well chosen, because that provision does not express a principle in Alexy's sense:

"[T]he example does *not* refute Alexy's theory, since art. 14 of the Spanish Constitution is a (vague) rule and not a principle in Alexy's sense. Besides, provisions such as Ch. 1 Sec. 2 of the Swedish Constitution, which are principles in that sense, *can* be complied with in different degrees." (Peczenik 1992, 330)

Actually, it is not easy to reply to this argument. This is so because, first of all, Peczenik does not explain why art. 14 of the Spanish Constitution is not a principle in Alexy's sense and what relevant differences there are, in that respect, between art. 14 of the Spanish Constitution and the provision of the Swedish Constitution he mentions. That provision stipulates that "public power must be exercised, respecting the equal value of all human beings, and the liberty and dignity of each individual person". Why should one agree with Peczenik that this provision expresses three principles — the equality, liberty and dignity of all individuals —, but deny that art. 14 of the Spanish Constitution expresses the principle of equality? In the provision of the Swedish Constitution just quoted as well as in art. 14 of the Spanish Constitution, there is that 'one-to-one link' to the corresponding values which Peczenik considers characteristic of principles. And neither one of the two provisions determines its generic conditions of application. So what is the difference between them that would allow us to hold that the first expresses three principles, and the second none? Besides, Peczenik asserts that the principles contained in the Swedish Constitution can be complied with in different degrees. But this assertion means that he takes for granted precisely what should be argued for: the possibility for each one of the three principles to be complied with by degree.

Peczenik's second argument in defence of Alexy's conception can be found in the next paragraph, where, after quoting our assertion that principles have greater 'expansive force' than rules, he writes:

"Without trying to explain the term 'expansive force', it can be pointed out that the main source of the justificatory force of principles consists in their one-to-one link to the corresponding values. Every principle corresponds to some value; it stipulates, for example, that equality, liberty and dignity are valuable. [...] A value can be defined as a criterion of evaluation. Each criterion can be satisfied to a greater or lesser degree [...] Each principle requires that the value it corresponds to should be respected as much as possible. But if that is so, then the possibility of complying with principles in different, greater or lesser, degrees is the most essential property of principles. And Alexy is right, whereas the critique of his theory by Atienza and Manero does not go deep enough." (Peczenik 1992, 330)

We basically agree with Peczenik that what he calls their "one-to-one link to the corresponding values" is characteristic of principles in the strict sense.[6] In fact, in our contribution we characterized principles in the strict sense in very similar terms. But from here, one cannot infer what Peczenik claims to infer. Of course, principles are criteria for the evaluation of behaviour. But it is simply wrong to say that criteria of evaluation can always be satisfied in degrees. The best example against this can be found in those legal standards we have called *rules* and which, for those who accept them, are also criteria for the evaluation of actions. Now, it is undisputed that rules do not admit of compliance in degrees: very simply, they are either complied with or not complied with. The same is true, in our opinion, with principles in the strict sense; and in Peczenik we do not find reasons that would make us change our mind on this.

Finally, to conclude, just a few words about what we have called the 'expansive force' of principles and which, in fact, we have sketched only very roughly in the text.[7] With respect to principles in the strict sense, their 'expansive force', for example, translates into the generation of rules that determine their prevalence (or non-prevalence) in some generic case. Let us return to Prieto's example of the will. If the court that must judge the case determines the prevalence of the principle of equality (or, more precisely, the principle of the prohibition of discrimination on grounds of conscience) over that of autonomy of the will, that determination translates into the construction of a rule, as the foundation of its decision, stipulating that under certain generic conditions (the making of a will, the conclusion of some private legal act) it is prohibited to discriminate on grounds of conscience. And if the *rationes decidendi* of the court judging the case are binding for other judicial organs, that rule becomes a rule of the corresponding legal system.

---

[6] On the connexion between principles and values, cf. below, ch. V.
[7] Ch. I, 3.2.1.

CHAPTER II
POWER-CONFERRING RULES

1.   Introduction

In this chapter, we will try to grapple with one of the most controversial questions in contemporary legal theory: the question about the nature of the norms that confer normative powers or, in a terminology that is more common among legal scholars, of the norms that confer competences (in public law) or capacities (in private law).

As is well-known, one of the central pillars of Hart's *The Concept of Law* is that power-conferring rules are seen as statements which cannot be reduced to mandatory norms. Only if we understand power-conferring rules as a special kind of normative statements, according to Hart, we can account for "the distinctive characteristics of law and of the activities possible within its framework" (Hart 1994, 41). Such rules, Hart says, "do not impose duties or obligations" (Hart 1994, 27); rather, they establish how one must proceed in order to bring about certain normative changes; for their addressees, they are "more like *instructions* how to bring about certain results than mandatory impositions of duty." (Hart 1982b, 219); "[t]hey appear then as an additional element introduced by the law into social life over and above that of coercive control" (Hart 1994, 41). All these and many other similar passages to be found in Hart's work suggest that power-conferring rules differ from mandatory norms with respect to their structure, the way in which they contribute to their addressees' practical reasoning, and their impact on social life. However, Hart's work does not offer a fully developed theory of power-conferring rules that would allow us precisely to account for them — and for their difference from mandatory norms — with respect to these three perspectives. Besides, in *The Concept of Law* the distinction between mandatory norms (or 'duty-imposing' norms) and power-conferring rules seems to be wrongly presented as similar to other, not equivalent distinctions. Thus, Hart uses three different criteria for distinguishing 'primary norms' from 'secondary norms': the first refers to the difference between duty-imposing rules and power-conferring rules; the second to that between rules regulating actions which imply movements or physical changes and rules concerning acts which lead to normative changes; and the third to that between rules about actions individuals should or should not do and rules about rules of the first type. Hart presents these three criteria as interchangeable, that is, as obviously having different connotations, but leading to the same classificatory results. But that is clearly not so. To give just one example: The provision contained in art. 53.1 of the Spanish Constitution ordering the legislator to respect the 'essential content' of certain rights and freedoms is a primary rule according to the first criterion, and a secondary rule according to the other two.[1]

2. What power-conferring rules are not

We will approach our topic by, first of all, excluding a few things, that is, we will begin by explaining what, in our view, power-conferring rules *are not*.

The first of these exclusions is the following: power-conferring rules are *not deontic or regulative norms*. This expression — 'deontic or regulative norms' — may sound strange and seems to be a perfect example of a pleonasm: if a norm is a sentence that commands, prohibits or permits something, then to speak of 'deontic norms' or 'regulative norms' is like speaking of 'rectangular rectangles'. The same thing happens if norms are understood as the meaning of such sentences. And it stays the same if norms are conceived as the result of acts of prescription (that is, of an order, a prohibition or a permission). Thus, whatever perspective (syntactic, semantic or pragmatic) we adopt to account for norms, their characteristic is always precisely their regulative or deontic character. Hence, if one assumes that this character is a defining element of norms, our thesis would be that power-conferring rules are not norms. Later, we will discuss the question of whether it makes sense to continue calling those rules, which — according to our thesis — do not order, prohibit or permit anything, norms, thus stretching the usual meaning of the word (or, more precisely, its intension — since power-conferring rules in our view have no deontic or regulative character —, but not its extension — since power-conferring rules usually belong to the denotation of the term 'norm').

---

[1] A detailed analysis of this problem and a proposal for the reconstruction of Hart's distinctions can be found in Ruiz Manero 1990, 99 ff. In any case, we should point out that whenever we use the expression 'secondary norms' in this book, we refer to rules concerning normative changes, whether by conferring the power to bring them about (power-conferring rules), by imposing duties with respect to the exercise of such powers (mandatory rules or principles regulating the exercise of normative powers), or merely by stipulating the states of affairs that determine the production of a normative change (purely constitutive rules which will be treated later).

Our second exclusion is the following: the thesis which presents power-conferring rules as *definitions, conceptual rules or qualifying dispositions* blurs important differences between such rules and two other kinds of sentences, namely, definitions, and what we will call 'purely constitutive rules'.

After these dismissals, we will now present our conception of power-conferring rules and explain how they differ, in our view, from deontic or regulative rules, from definitions, and from purely constitutive norms.

2.1. First exclusion: Power-conferring rules are not deontic or regulative norms

Since the understanding of power-conferring norms as deontic or regulative norms or, more precisely, as permissive norms is very widely shared (if only implicitly, in some cases), we will not attempt to mention all, or even what could be called a significant sample of, the authors holding this view. Instead, we will proceed as follows: We will show (following von Wright 1963 and Alchourrón/Bulygin 1971) that the conception of power-conferring rules as permissive norms cannot account for the *irregular use* of the respective powers; that if one tries to account for that irregular use on such a conception, as is the case with Kelsen's theory, the result is a totally distorted picture of a legal system, leading to the destruction from within of the theory itself; and, finally, that if a basically non-deontic definition of power-conferring rules is complemented (as in the case of MacCormick 1986) by a certain deontic surrounding, this only leads to confusion and misunderstandings.

2.1.1. Von Wright treats power-conferring rules in ch. X of *Norm and Action* (von Wright 1963) under the label of 'norms of higher order'. Norms of higher order are norms whose content are normative acts, where that means the acts of giving or cancelling prescriptions. Now, with the help of this concept of norms of higher order, von Wright writes, "we can illuminate one of the most controversial and debated notions of a theory of norms, *viz.* the notion of *validity*" (ibid., 194). "There are", von Wright continues, "at least *two* different, relevant meanings of the words 'valid' and 'validity' in connexion with norms. [...] One sense in which a norm can be said to be valid is [...] the factual aspect of law as the efficacy of a commanding will" (ibid., 194 ff.). This sense, then, is more commonly known today as simply the 'efficacy' or 'effectiveness' of a norm and, for our purposes, can be left aside. But besides this factual sense, norms are also said to be 'valid', von Wright goes on, "in the normative sense of 'legality'" (ibid., 196). This is the sense we are interested in, because it can illuminate the deficits of the conception of 'norms of higher order' or rules conferring normative powers as permissive norms. Let us take a closer look at von Wright's explication:

"Under this other meaning the validity of a norm means that the norm exists and that, in addition, there exists another norm which permitted the authority of the first norm to issue it. If we decide to call the act of issuing a norm *legal* (or lawful) when there is a norm permitting this act, then we may also say that *the validity*, in the sense now contemplated, *of a norm means the legality of the act of issuing this norm.*" (ibid., 195)

"A norm is valid when the act of issuing this norm is permitted. It is a theorem of deontic logic that, if an act is commanded, then it is also permitted. Therefore, an order to issue norms entails that the norms issued under that order are also valid, *i. e.* their issuing is permitted because commanded." (ibid., 198 f.)

"We could sharpen our definition of validity in such a way that to say that a norm is valid shall mean that the authority who issues it has a permission amounting to a right to issue the norm. Normative competence or power would then mean permissions in the stronger sense of rights to perform certain normative acts. [...] I think that this reshaping of our definitions should take place. The higher-order permissions of which we have here been talking should be regarded as rights." (ibid., 206)

According to von Wright, normative competence or power thus amounts to the permission (whether in the stronger sense of a 'right' or not) to perform certain normative acts, that is, certain acts of giving or cancelling prescriptions (of introducing or expelling norms). And a norm is valid — in the pertinent normative sense — if the act of issuing it is permitted by another higher-order norm.

Although they explicitly consider a somewhat more restricted domain — not that of competence in general, but only that of judicial competence —, Alchourrón and Bulygin, in *Normative Systems*, have advocated a conception of power-conferring rules that basically coincides with that of von Wright. They offer the following interpretation:

"Following traditional terminology, we shall call norms of competence the norms conferring jurisdictional powers on the judge, that is, the power to judge. Such norms establish that certain persons can (are competent to) deal with certain kinds of cases and issue certain kinds of decisions, observing certain formalities.
Norms of competence are norms of conduct for the judges, if we regard them as permissive norms establishing the permission to perform certain acts in certain circumstances. [...] It is important to stress the difference between the norms of competence (which are permissive) and

those which impose obligations and prohibitions on judges, which also are norms of conduct." (Alchourrón/Bulygin 1971, 151)

The conclusion from this is the same as in the case of von Wright: To say that "organ X is competent to perform the normative act Y" means exactly the same as to say "organ X has been permitted to perform the normative act Y".

Now, legal organs, e. g., legislative or jurisdictional organs, not only perform *regular* normative acts like, for example, issuing a law that does not violate the limits set by the constitution, or a judicial decision in accordance with the law. As everybody knows, sometimes they also perform *irregular* normative acts, like issuing an unconstitutional law, or a judicial decision *contra legem*. With respect to this — as long as we understand power-conferring rules as permissive norms — obviously there are only two alternatives: either the legal organs *are not permitted* to perform such normative acts; or *they are indeed permitted* to perform them. But, as we will presently see, this constitutes a real dilemma. Because if we lean towards the first horn and say that such normative acts are not permitted, then we cannot explain that such normative acts do have legal effects, that is, that the unconstitutional law is a *law* and the illegal judicial decision a *judicial decision*. The existence of a system of appeals is no argument against this, if only for two simple reasons: The first is, obviously, that the illegal judicial decision could come from, or be upheld by, the court that is, in the respective case, the last instance, just as an unconstitutional law may be beyond further revision, for example because the organ entrusted with controlling constitutionality has already decided so. The second is that the very fact that unconstitutional laws or illegal judicial decisions can be appealed shows that such laws or judicial decisions are precisely what they claim to be, that is, laws or judicial decisions. Because there is no appeal against a 'law' issued, let's say, by a group of professors, or a 'judicial decision' issued by an undergraduate student in an exercise class in criminal law. And the difference lies not only in the fact that parliament and the courts are legal organs, while professoral meetings and students are not; because legal organs themselves can sometimes be in exactly the same position: Nobody would say that what happens when a criminal court issues an illegal decision in a murder case falling under its jurisdiction is the same as what happens when the same court would decide, for example, to declare war on Serbia. In the first case, the court has issued a *judicial decision* whose execution, if it is not appealed, or if it is appealed, but upheld by the higher courts, is legally binding. In the second case (that of the declaration of war), in contrast, the decision of the court is legally just as irrelevant as the 'judicial decision' issued by the undergraduate. But if we say that the domain of competence of an organ coincides with the domain of the normative acts it is permitted to perform, then we cannot account for this crucial difference, because to issue judicial decisions *contra legem* is just as much outside of this range of permissions as to declare war on foreign powers.

Let us now see what happens if we tend towards the second alternative we have with respect to *irregular* normative acts if we understand power-conferring rules as permissive norms, that is, the alternative to say that such acts are permitted. This is Kelsen's position.

2.1.2. Our purpose here is not to undertake Kelsenian philology, or to follow the meanderings of his evolution concerning the higher norms that determine the production of lower norms.[2] We are only interested in his thesis that *irregular* normative acts are permitted acts or, more precisely, that there are no such irregular normative acts because the law permits its organs of production to issue norms of whatever content (we will consider this thesis in the version which is now commonly referred to as the 'classical' one, that is, that of the second edition of the *Pure Theory of Law*). Let us begin with Kelsen's statement of the problem:

"[S]ince a norm belongs to a certain legal order only because and so far as it is in accord with the higher norm that determines its creation, the problem arises of a possible conflict between a higher and a lower norm. The question then arises: What is the law, if a norm is not in conformity with the norm that prescribes its creation and, especially, if it is not in conformity with the norm predetermining its content. [...] a legal norm which might be said to be in conflict with the norm that determines its creation could not be regarded as a valid legal norm — it would be null, which means it would not be a legal norm at all." (Kelsen 1967, 267)

Anyone would say that what follows from this is that, since judicial decisions *contra legem* and unconstitutional laws are, by definition, norms which are not in accord with the higher norms determining their production, such decisions and laws are not legal norms at all. However, this is not what Kelsen thinks. He wants to account for the fact that such laws or judicial decisions can either be *appealed* — if there is an organ entrusted with controlling the constitutionality of laws, in the first case, or if they come from a court that is not the last

---

[2] Cf. on this, with respect to the 'classical' Kelsen of 1960 as well as to his later work, Ruiz Manero 1990, 51 ff.; also, Ruiz Manero 1992 and 1994.

instance, in the second — or are *definitely binding* — if there is no such organ, or this organ cannot be appealed to, in the first case, or if they come from a court of last instance, in the second. But the fact that they can be appealed in some cases as well as the fact that they are definitely binding in others both *imply that the law itself recognizes them as law*. The problem then is how to reconcile this fact with the thesis that a norm belongs to a legal order only "because and so far as it is in accord with the higher norm that determines its creation ". Kelsen's solution is to reformulate those "higher norms" in such a way that they permit the issuing of norms of any content whatsoever. Let us illustrate this with some quotes:

"[N]ot only is a general norm valid that predetermines the content of the judicial decision, but also a general norm according to which the court may itself determine the content of the individual norm to be created by the court. The two norms form a unit [...]" (Kelsen 1967, 269)

"The meaning of the constitution regulating legislation is not that valid statutes may come into being only in the way directly stipulated by the constitution, but also in a way determined by the legislative organ itself. The provisions of the constitution which regulate legislation have an alternative character." (Kelsen 1967, 273)[3]

To sum up, all "higher norms" that "determine" the creation of norms — in Kelsen's conception, all legal norms except individual ones (like judicial decisions or administrative resolutions) that order specific acts of material execution — are said to have the form of a disjunction, consisting in their explicit content and a *tacit alternative clause* that permits the respective norm-creating organ to disregard that explicit content. In this way, all normative acts are permitted by higher-order norms, and all norms issued through such acts are "in accord" with those higher norms, just as any state of affairs in the world is 'in accord' with any tautology. But just as a tautology does not have any informative content, a legal system composed — with the one exception indicated — entirely of norms with a tautological content (that is, norms that cannot by any

---

[3] The German original is more extensive and illuminating on this point: "Das bedeutet aber, daß der Sinn der die Gesetzgebung regelnden Normen der Verfassung nicht der ist, daß gültige Gesetze nur auf die durch die Verfassung direkt bestimmte Weise, sondern daß sie auch auf eine andere, durch das Gesetzgebungsorgan selbst zu bestimmende Weise zustandekommen können. Die Verfassung ermächtigt den Gesetzgeber, auch in einem anderen Verfahren als jenem, das durch die Normen der Verfassung direkt bestimmt ist, generelle Rechtsnormen zu erzeugen und diesen Normen auch einen anderen Inhalt zu geben als jenen, den die Normen der Verfassung direkt bestimmen. [...] Die die Gesetzgebung regelnden Bestimmungen der Verfassung haben den Charakter von Alternativbestimmungen" (Kelsen 1960, 277).

means be violated) does have no normative significance — that means, it cannot serve as a guide, or a criterion of evaluation, or a model of interpretation — concerning the conduct of its organs, since whatever that conduct may be, it is necessarily in accord with the system.

Kelsen's attempt to account for irregular normative acts, based on a conception of power-conferring norms as permissive norms, thus leads to a form of dissolution of the law as a normative system that regulates its own creation.

2.1.3. Therefore, it seems that any attempt to define what it is to have the normative power to perform a normative act Y in terms of having the permission to perform that normative act is bound to lead to a dilemma whose two horns are both dead ends. Neil MacCormick, in his famous article 'Law and Institutional Fact' (1986), has proposed a characterization of power-conferring rules in non-deontic terms which, in our view, points into the right direction. But by adding a certain deontic surrounding to his initial definition, MacCormick's proposal, as we already anticipated and will now show, is led back into the messy terrain it had been saved from by the non-deontic character of its point of departure.

In 'Law as Institutional Fact', MacCormick presents legal institutions (legislation, contracts of commercial exchange, wills) as consisting of three kinds of rules: 'institutive' rules determining the essential conditions for the existence of a particular case (the Law of University Reform, the marriage between Peter and Mary) of a legal institution (legislation, marriage); 'consequential' rules determining the legal consequences of the existence of a particular instance of an institution; and 'terminative' rules determining when a particular case of an institution ceases to exist.

Power-conferring rules can be of any one of these three kinds of rules, although not all the rules of each of these three kinds are power-conferring rules. MacCormick nevertheless limits his analysis of power-conferring rules to 'institutive' rules. An institutive rule that confers a legal power, in his view, has the following 'general form':

"If a person having qualifications q performs act a by procedure p and if the circumstances are c, then a valid instance of the institution I exists." (MacCormick 1986, 65)

A formulation of this type — or the more detailed one to be found in MacCormick (1993) — seems to have the advantage of enabling us to avoid the dilemma which, with respect to the irregular exercise of normative powers, arises, as we have seen, for whoever insists on understanding power-conferring rules as permissive norms. Since it apparently does not contain any deontic element, MacCormick's formulation of a power-conferring rule is compatible with any normative qualification, by any other rule, of the conduct consisting in the performance of the normative act *a*. If that act, for example, consists in issuing a judicial decision contrary to law, performance of that act is prohibited by a legal rule, but that does not mean that the competent judge cannot successfully use the rule that confers on him the power to issue a decision, or that his decision *contra legem* isn't a *valid* (that is, recognized as such by the law) particular case of the legal institution 'judicial decision'.

Unfortunately, MacCormick does not take this — in our view, promising — way he himself opens up with his general formulation of a power-conferring rule. The reason for this is that he does not understand 'a *valid* particular instance of institution *I*' to be synonymous with 'a particular instance of institution *I*, *recognized as such by the law*', but with 'a particular instance of institution *I* which *organs of control do not have the duty to cancel*'. Thus, the deontic element that seemed to be absent in his general formulation reappears. Let's look at it somewhat more closely. MacCormick writes that we cannot be sure about the conditions that, in each particular case, are *sufficient* for the validity of a particular instance of a legal institution; as an example, he refers to the will in *Riggs v. Palmer*, the case made popular by Dworkin. As is well-known, that will, which favored the grandson of the testator, had been made according to all the requirements of the corresponding New York statute. However, as is also well-known, the court ruled that the will should be annulled, on the grounds of the principle that nobody shall benefit from his own illicit action, because the grandson himself had murdered the testator. MacCormick adds that similar situations can be found: "To take but one branch of law, administrative law abounds with illustrations of cases in which decisions made by competent bodies in accordance with all expressed statutory requirements have been set aside for some defect in the circumstances or manner of making the decision, the justification given by the court being in terms of appeal to general legal principle" (MacCormick 1986, 70). And he concludes: "It is the open-ended nature of the exceptions justified by the principles of natural justice, abuse of discretion, and such like, which would be fatal to any attempt to represent the express institutive rules as containing necessary and sufficient conditions for valid adjudication by tribunals or whatever" (ibid., 70).

As we already said, in our view MacCormick here confuses two quite different things, both entailed by the ambiguous expression 'valid'. Because one can speak of a 'valid' will, a 'valid' administrative resolution, a 'valid' judicial decision, etc., simply to say 'recognized as such' — that is, as a will, an administrative resolution, or a judicial decision — 'by the law'. But we can also speak of a 'valid' will, administrative resolution, or judicial decision in order to say that the respective act does not violate any of the requirements violation of which generates a duty of annulation on the part of some organ of control. MacCormick's examples of a will, an administrative resolution or the exercise of judicial power are obviously 'valid' in the first sense, and that is precisely a necessary condition for their being 'invalid' in the second sense (that is, that they can and ought to be annulled). In this context, of course, it does not matter whether they should be annulled because they contradict a general legal principle or because they contradict some rather specific rule. What matters is that if a will, an administrative resolution or a judicial decision should be annulled for violating some mandatory legal norm (whether that is a rule or a principle) this implies that the thing to be annulled *is* (recognized by the law as) a will, an administrative resolution or a judicial decision. And the problem a theory of power-conferring rules must solve is to spell out the conditions for successfully (that is, in a way that is recognized by the law) creating things like laws, wills, resolutions or judicial decisions — whether or not those laws, wills, etc., may then be annulled for violating some mandatory legal norm (again, we insist, irrespective of whether this is a specific rule or a general principle).

2.2. Second exclusion: Power-conferring rules cannot adequately be understood in terms of definitions, conceptual rules, or qualifying dispositions

In several publications after *Normative Systems*, Carlos Alchourrón and Eugenio Bulygin have defended the view that power-conferring rules are definitions or conceptual rules. And Rafael Hernández Marín understands them to be part of what he calls 'qualifying dispositions'. In our opinion, these are simply two different ways of formulating the same thesis. Alchourrón and Bulygin as well as Hernández Marín think that the law consists of two kinds of provisions (although they concede the possibility — which none of them elaborates — that there may be legal provisions of other types): norms of conduct, on the one hand, and definitions or conceptual rules (Alchourrón/ Bulygin) or qualifying dispositions (Hernández Marín), on the other. In our view, Alchourrón and Bulygin's concept of 'definition' or 'conceptual rule' is (at least as far as it concerns the problem we are interested in here) the same concept Hernández Marín designates with the expression 'qualifying disposition'. This will become clear when we look at their own formulations. Let us begin with Hernández Marín:

"In the law, there are dispositions of two different kinds: norms in the strict sense, i. e., norms of obligation (or prohibition), and qualifying norms. Perhaps there are also other types of dispositions, but here we will only consider norms and qualifying dispositions." (Hernández Marín 1984, 29)

"Qualifying dispositions are sentences attributing a property to, or including in a certain class, all entities with a certain property or belonging to a certain class." (Ibid., 30 f.)

Examples of such qualifying dispositions are dispositions about the ways in which property can be acquired, i. e., "dispositions qualifying the buyer, the heir, etc., of a thing as the owner of that thing"; "the disposition by which someone is appointed minister (mailman, professor, etc.) [which] qualifies as a minister (mailman, professor, etc.) all entities equal to that someone"; "dispositions by which it is defined what an official document is [which] qualify certain objects as official documents" (ibid., 31). The same is true for rules conferring public or private normative powers:

"Dispositions 'conceding private powers' (H. L. A. Hart) also are qualifying dispositions since they qualify as a will, a marriage, a sale, etc., acts performed by persons who satisfy certain requirements." (Ibid., 31)

"Norms of competence are qualifying dispositions. [...] Because, in fact, a norm or legal disposition of competence is a legal disposition qualifying as legal, or as belonging to the law, all sentences with a certain property: that they come from organ O, are in accordance with procedure P and on subject S. Hence, what a norm of competence qualifies are sentences or dispositions; and the qualification given by a legal norm of competence to the dispositions it refers to is that of being valid, or legal, or of belonging to the law. I must confess that I do not know of any other acceptable meaning of legal validity than that of belonging to the law, and in that sense, the expressions 'validity' and 'valid' are superfluous." (Ibid., 38 and 40)

"Most important are the consequences of that interpretation of norms of competence. One of them [...] is that norms of competence cannot be violated. And for that same reason, they also cannot be obeyed, they cannot be effective. As a consequence, since the application of the law is a 'form' of obeying the law or of legal norms being effective, norms of competence cannot be applied. Hence, it is not true that when a legal norm is created another legal norm — a legal norm of competence — is applied." (Ibid., 42; the same theses can be found in a more succinct form in Hernández Marín 1989, 160 ff.)

Let us now compare this with some passages from Alchourrón and Bulygin:

"Definitions serve to identify the norms that contain the defined terms, and this is the only function of definitions." (Alchourrón/Bulygin 1991a, 449)

"While in a norm certain words are *used* for referring to certain conducts in order to regulate or permit them, that is, to declare them obligatory, prohibited, or permitted, in a definition certain words are used in order to indicate the sense of other words that are *mentioned*, but not used." (Ibid., 455)

"Against those authors who, in order to preserve the idea that all legal norms are of the same kind (norms of obligation), assimilate nullity to a sanction, Hart argues that these are two completely different notions. Hart's aim is to show that in the law there are two different types of rules which he calls (primary) rules of obligation and (secondary) power-conferring rules [...] Hart has hit upon a very important point here. Sanction and nullity are indeed — as Hart's argument shows — two distinct notions not reducible to each other. The different ways in which a sanction and nullity work attest to a radical difference between the two kinds of rules. Power-conferring rules are indeed different from rules of obligation, but one can ask the question: What are those rules? [...] Hart's argument clearly shows that power-conferring rules cannot be interpreted simply as permissive norms, because it does not make sense to speak of nullity in the case of a permissive norm. If someone does not make use of an authorization, he does not thereby perform an act that is null. In contrast, if there is a definition that stipulates what requirements an act, a document, or a norm must satisfy, absence of any one of the essential requirements will determine the nullity of the act, document, or norm [...]. The presence of both these institutions — sanctions and nullity — is a clear sign that there exist in the law two radically different kinds of rules: norms of conduct, on the one hand, and conceptual rules or definitions, on the other." (Ibid., 461 ff.)

"The existence of these two kinds of institutions, sanctions and nullity, is a sign that there exist in the law (at least) two kinds of rules: norms of conduct and conceptual rules, or, in Searle's terminology, regulative and constitutive rules. [...] I think that the distinction between conceptual rules and rules of conduct or — what amounts to the same — between constitutive and regulative rules offers an interesting conceptual tool for an explication of the obscure concept of legal power (rechtliches Können). Its explication in terms of permission is rather problematic [...] It seems more promising to try to explain it in terms of conceptual or constitutive rules. On that interpretation, the norms that establish the competence of the legislator (in its personal, material, and procedural aspects) define the concept of legislator and make the activity of legislating possible. [...] [I]t is not always easy to find out whether we are dealing with a (defining or constitutive) norm of competence or with a norm of conduct. I think one of the criteria could be the following: When the 'legal power' persists despite of the prohibition to exercise it [...] we are dealing with competence; when the prohibition causes the 'legal power' to disappear, we are dealing with a permission, a liberty, or a privilege conferred by a norm of conduct and not by a conceptual rule." (Bulygin 1991, 496 f.)

In our view, the characterization of power-conferring rules in terms of conceptual rules, definitions or qualifying dispositions does have advantages over their characterization in deontic terms. For the latter, the *irregular* exercise of normative powers was the crucial test that determined its failure. This is not the case with the position we are considering now. As the last quote from Bulygin shows, this conception is perfectly able to account for the fact that the domain of what a subject X can do, *in the sense of having the normative power to do it*, may not coincide with the domain of what that same subject can do, *in the sense of being permitted to do it*.

The problems with this position rather seem to be that it puts into one and the same category — that of definitions, conceptual or constitutive rules, or qualifying dispositions — legal provisions sufficiently different in relevant aspects as to require the elaboration of separate categories.[4] Consider, e. g., the following two examples (both would be conceptual rules, definitions or constitutive rules according to Alchourrón/Bulygin, or qualifying dispositions according to Hernández Marín):

1) "To the effects of the present law, 'rural estates' will be understood to mean estates with the following characteristics ..."

2) "In order validly to lay down a will, the presence of two witnesses is required."

Apparently, the function of provision 1) is exclusively that of identifying the norms expressed by the norm-formulations which in the corresponding law use the term 'rural estate', while provision 2), besides the function of stipulating a necessary condition for certain expressions of will to be identifiable as 'a will', also has the function of indicating what (or part of what) someone who wishes to generate, through his or her action, all the normative consequences other norms link to the valid giving of a will, must do. That means that with the use of provision 1) all one can do is identify the norms contained in the respective law, while with the use of provision 2), one can do more than just identify a will, namely, one can *make* a will.

---

[4] A similar, though somewhat differently developed argument can be found in Aguiló Regla (1990).

3. What power-conferring rules are

3.1. Three approaches and some ontological assumptions

Let us now move to the constructive part. On the basis of a pattern that parallels the one used in the previous chapter with respect to mandatory norms, in what follows we will try to clarify from three different angles what power-conferring rules are. From the first one — which can be called the *structural* point of view — we will try to see what elements a power-conferring rule consists of and how they interrelate, and how such rules differ, in that respect, from deontic (or regulative) norms and from definitions; from the second perspective — which may be called the *functional* or *justificatory* one — we will consider the different roles power-conferring rules have, compared with regulative rules or definitions, in practical-legal reasoning; and finally — from a perspective which we can call the *social* one — we will formulate some observations about how power-conferring rules are related to power in the non-normative sense and to interests.

Before we present our view of the structure of power-conferring rules and the way in which they come into play in practical reasoning, we think it is necessary to explicate some ontological assumptions which are at the foundation of our answer to those questions. They are very simple — although probably not exactly obvious — and consist in distinguishing between facts and actions which, in turn, can be generic or individual, on the one hand, and natural or institutional, on the other. For example, to turn 18 years old is a generic fact; the fact that X turns 18 today (on June 30, 1993) is an individual fact. To commit murder is a generic action; the murder committed by ETA last week in Madrid is an individual action. It must be emphasized that the qualification of — generic or individual — facts and actions takes place on the level of language; thus, for example, one and the same body movements can be presented as different facts or actions: as shooting, killing, committing murder, etc. The distinction between natural (non-institutional) facts or actions and institutional facts or actions operates on this level. By natural (non-institutional) facts or actions we understand those stated independently of rules belonging to some institution (for our purposes, to some positive law).[5] For example, to turn 18 is

---

[5] This means that, depending on the respective institution, one and the same action can appear as a natural or as an institutional action. For instance, from the perspective of the social institution we call 'law', 'promising' is a natural action, while 'concluding a legal treaty' is an institutional

a natural fact since we can say that a persons turns 18 today independently of what the rules of some positive law may or may not stipulate. In contrast, to come 'of age' is an institutional fact since when we say that someone is of age we are — at least in part — using a rule (in a sense that also includes sentences like definitions) of some positive law. Similarly, to kill someone is a natural action, whereas to commit murder (in the sense of art. 405 of the Spanish Criminal Code) is an institutional action. Note that our distinction between the natural (or non-institutional) and the institutional commits us to very little; particularly, since it operates on the linguistic level, it is immune to criticisms like Joseph Raz's against Searle's theory of constitutive rules[6], and the distinction thus drawn also does not mean to assert or to deny that there is something like brute facts or actions, if by this we understand facts or actions that can be stated independently of any interpretational scheme whatsoever.

A state of affairs is a set of natural and/or institutional facts which, in turn, may or may not be the result of natural or institutional actions. For example, to be of age and married is a set of two institutional facts the second of which, in turn, is the result of an institutional action.

3.2. A structural approach

3.2.1. If we suppose — as we think is correct — that all general legal norms follow a conditional pattern in which one can therefore distinguish an antecedent and a consequent, then we can say that the antecedent of a general norm always refers to more or less complex states of affairs. As for the norms we have called deontic or regulative, their antecedent consists in a state of affairs that may or may not contain institutional facts (that is, regulative norms can regulate natural or institutional states of affairs), and their consequent is formed by a natural or institutional action or state of affairs and a deontic operator. Their canonical form is as follows: 'If state of affairs X obtains, then action Y ought to be done by (is obligatory, prohibited, or permitted for) Z' or 'If state of affairs X obtains, then it is obligatory, prohibited, or permitted for Z to attain (or try to attain) end (state of affairs) F.' For example, 'it is prohibited to kill another person' (which presupposes the existence of a certain state of affairs, namely, that the other person is alive, that there is an opportunity to kill her, etc.) or 'a person that has committed manslaughter (the state of affairs consist-

---

action; from the perspective of social morality, however, 'promising' is an institutional action, and 'to say the words "I promise"' is a natural action.

[6] Cf. Raz (1990), 108-111, and, with respect to this, González Lagier (1993).

ing in having performed the action of taking the life of another person) shall be sentenced by the judge to a minor term in prison'.[7] In other words, deontic or regulative norms can be *primary* or *secondary* norms, that is, their deontic operator can work on a *natural* or on an *institutional* action or state of affairs. Now, power-conferring rules have a very different structure. Their antecedent consists of two elements: a state of affairs containing either natural or institutional facts, and an action that can also be either natural or institutional; and their consequent does not consist of an action modalized by a deontic operator (a solution), but of a kind of institutional fact which we will call an institutional or normative result (generally, a result is a change in a state of affairs produced by an action or a fact, and therefore one can speak of normative results — e. g., 'to have assumed an obligation' — and of non-normative results — 'to have killed a person', 'to have died'). The canonical form of a power-conferring rule would then be as follows: 'If state of affairs X obtains and Z performs action Y, then institutional result (or normative change) R is produced.'[8] For instance, persons of age can validly get married, that is, if they perform certain actions (consisting in filling in a number of forms, stating their desire to be married before a judge or another authority, etc.) an institutional result is produced which consists in a modification of their normative status: they acquire certain rights and obligations. Let us now analyze somewhat more closely the differences between the two types of norms.

3.2.2. The first and most obvious one is that, according to what has just been said, regulative norms are (expressed in) deontic sentences, but this is not the case with power-conferring rules: in what we have called their 'canonical form', no deontic operator appears. Someone could say that this neglects a fundamental aspect of those rules. For example, in the case of the rule conferring the power to get married, it is a fundamental fact that getting married is an institutional result that is facultative:[9] it is just as permitted to bring it about as not to do so. Now, in our opinion, this possible objection can be answered by saying

---

[7] [*Translator's note:* A 'minor prison term', in Spanish law, is a sentence between 12 years and one day and 20 years.]

[8] The reference to action Y is an abbreviation for what is normally either a conjunction of actions (a course of action), a disjunction of actions or courses of action, or a combination of both. On this point, see part 2.5 of the appendix to the present chapter.

[9] Here and in what follows, we use the term 'facultative' for referring to acts that are, so to speak, 'deontically optional', in order to distinguish them from acts which are 'anankastically optional' and which we will simply refer to as 'optional'.

that the rule conferring the power to get married is one thing, and the (regulative) norm stipulating that in state of affairs *X* (simplifying somewhat: if one is single and of age) getting married is facultative is quite another. The same institutional result is, on the one hand, the consequent of a power-conferring rule — indicating how to bring it about — and, on the other, it also is (modalized by the deontic operator 'facultative') the consequent of a regulative norm. In other words: one thing is to confer a (normative) power and another to regulate (as facultative, obligatory, or prohibited) the exercise of that power. And it is precisely their being two different things what explains that a judge, the administration or the legislator are successful (that is, that they produce the intended normative result, irrespective of its possible posterior annulation by some organ of control) when they issue a judicial decision *contra legem*, an illegal regulation, or an unconstitutional law. Of course, one can always say that the norm conferring a power and the regulative norm that deontically modalizes the exercise of that power can be seen as a functional unity, but that does not preclude that they can — and should — be kept apart for analytical purposes, just as we distinguish, for example, between the regulative norm that punishes murder with a major prison term and the definition stipulating how much a major prison term is.

A second objection is that, from the point of view of a *bad man* (in Holmes' sense) who pursues somewhat strange ends, some rules that look like an unquestionable example of regulative rules to us may appear (or be used) as power-conferring rules. Neil MacCormick (1981, 75) gives the example of a literary character — W. H. Davis' 'Super-Tramp' — who knows that under certain conditions a certain behaviour is a crime, and since he wants to commit a crime in order to incur a sanction and be comfortably installed in prison for a good while, he performs such an action (or, in terms of an institutional action, he commits the crime in question). It would be strange, however, to say that that man has used a rule that confers on him the power to change his own normative situation, by making him liable to a sanction. MacCormick's answer in order to circumvent this difficulty consists in saying that "[p]ower is conferred by a rule when the rule contains a condition which is satisfied only by an act performed with the (actual or imputed) intention of invoking the rule." (ibid., 74). The notion of 'invoking a rule' as it is developed by MacCormick seems somewhat obscure. Besides, we think that the problem does have a less sophisticated solution. However, it forces us — as is often the case with the key questions of legal theory — to look *away from the norms* and turn to the *purpose generally attributed to each one of them* (in our judgment, this is also one of the main lessons to be learned from Hart's work).[10] With respect to this, it seems that nobody would hold that the purpose of the norms of criminal law is to enable people to change their normative status by performing the actions corresponding to the institutional qualification of 'crimes', but rather to deter people from performing such actions. Of course, the use of this criterion may leave some cases in the shadow; but far from being particular of this case, that is something that also happens with other distinctions — as the one between a fine and a tax — which apparently can only be drawn by referring to the purpose — whether prohibitive or not — generally attributed to the rule in question (on this last point, cf. Hart 1983b, 295 ff.).

3.2.3. Since this structural characterization of power-conferring rules denies that they are deontic or regulative rules, we must explain precisely in what sense they are still norms or, if you prefer, what kind of norms they are. Our answer is that they are anankastic-constitutive rules that can be used as institutional technical rules.

Natural technical rules are based on anankastic propositions stipulating that, given some state of affairs X (e. g., the existence of water in a vessel), if action Y is performed (if it is heated to $100°$ C) R will result (the water will boil). From this follows the technical rule that if you want to boil water you must heat it to $100°$ C. In the example of the rule conferring the power to get married, the *constitutive* aspect derives from the fact that it is the legislator who — by issuing the rule — constitutes the (institutional, not natural) state of affairs consisting in getting married. But apart from this important difference, the legislator's rule also leads to a technical rule which stipulates that, given some state of affairs X, if Z wants to obtain some specific result (bring about a new — institutional — state of affairs, R) she must use the power conferred on her, that is, she must perform action Y. In both cases, one could make a distinction according to whether the antecedent of the conditional is a necessary, a sufficient, or a necessary and sufficient condition of the consequent (the result).[11]

---

[10] Raz (1990, 103) has expressed the same idea in the following words: "An act is the exercise of a normative power if, and only if, it is recognized as effecting a normative change because, among other possible justifications, it is an act of a type such that it is reasonable to expect that, if recognized as effecting a normative change, acts of this type will be generally performed only if the persons concerned want to secure this normative change."

[11] Amedeo Conte (1985a and 1985b) and, following him, Gianpaolo Azzoni (1988) have used the concept of an anankastic-constitutive rule in the sense of a rule that establishes a *necessary*

Thus, we can say that a complete power-conferring rule is one that stipulates the sufficient (or necessary and sufficient) conditions for attaining the result, and that a norm merely containing some of the necessary conditions is only an incomplete norm.

3.2.4. Let us now look at the distinction the structural point of view sees between power-conferring rules and other kinds of legal statements. In our opinion, power-conferring rules should be distinguished not only from definitions but also from what we call purely constitutive rules. We think that in the article by Alchourrón and Bulygin (1991a) mentioned above, two parts can be distinguished. In the first part, they characterize definitions as sentences that do not express norms (although they may have normative consequences), but rather allow us to identify norms by elucidating the sense in which certain expressions are used (it should be remembered that for Alchourrón and Bulygin — at least in that paper — a norm is the *meaning* of a sentence, not a sentence itself). The canonical form of a definition (and they make no distinction between legislative and private, or unofficial, definitions) is the following: "'...' means ...",

---

condition for that which it regulates, as, e. g., in the rule that stipulates that a holograph will must be completely handwritten by the testator. As will have been noticed, we use the term 'anankastic-constitutive rule' here for referring to power-conferring rules insofar as they stipulate either a necessary, or a sufficient, or a necessary and sufficient condition for some institutional result.

In the case of Azzoni, the rules setting a necessary condition (anankastic-constitutive rules), a sufficient condition (metatetic-constitutive rules), or a necessary and sufficient condition (nomic-constitutive rules) are subspecies of the general type of 'hypothetical-constitutive rules'. They do not, however, exhaust the still more general category of 'constitutive rules', which also includes those rules that are a *necessary* condition (eidetic-constitutive rules), a *sufficient* condition (thetic-constitutive rules), or a necessary and sufficient condition (noetic-constitutive rules) of that which they are rules of. From all these types, one must distinguish hypothetical rules, i. e., rules that *presuppose* necessary conditions (anankastic rules), sufficient conditions (metatetic rules), or necessary and sufficient conditions (nomic rules) of that which they are rules of. The concept of a 'hypothetical rule', in turn, basically coincides with that of a 'technical rule', although it does not entirely exhaust it.

The fact that the term 'constitutive rule' can have such a plurality of meanings will help the reader understand why we did not adopt as our starting point one of the categories of 'constitutive rules' that can be found in the literature. Riccardo Guastini has studied the different concepts of constitutive rule authors like Searle, Ross and Carcaterra work with and has shown — in our view, convincingly — that the concept of 'constitutive rule' has been ambiguous from its very beginning in the work of Searle (Guastini 1983 and 1990).

As Azzoni has correctly pointed out, the question about what a constitutive rule is "is a wrong question, because it starts from a wrong assumption, namely, that the term 'constitutive rule' is unambiguous" (Azzoni 1988, 2). Later, we will ourselves speak of "purely constitutive rules" in order to refer to rules differing from power-conferring (anankastic-constitutive) rules in that no action must be performed for the institutional results mentioned in them to come about.

where "'...'" stands for the expression to be defined *(definiendum)*, and "..." for the words used to convey its meaning *(definiens)*. For example: "'To be of age' means to be at least 18 years old." In the second part, Alchourrón and Bulygin hold that a legal order can be constructed satisfactorily with only two categories of sentences, namely, norms of conduct and definitions or conceptual rules. As we have already seen, this leads them to characterize power-conferring rules as definitions.

Now, while the first thesis seems to be a happy characterization of a definition, we believe that the second one is false. The sentence saying "Heir is the person entitled to inherit an estate as a whole, legatee the person entitled to a partial inheritance" (art. 660 of the Spanish Civil Code) is, in fact, a definition, since it only specifies the sense in which the terms 'heir' and 'legatee' are used. But other sentences Alchourrón/Bulygin regard as definitions are, in our view, ambiguous. This is the case, for instance, with the sentence stipulating that one comes of age with one's 18th birthday. It can, of course, be interpreted as a definition: 'of age' means 'being at least 18 years old'. But it can also be understood as a sentence stipulating that the production of a certain state of affairs (having celebrated one's 18th birthday) determines the production of a normative change: that one reaches the normative status of 'being of age'. And this is also true — now without any ambiguity at all — with the sentence "A person's rights of inheritance are transferred at the moment of death" (art. 657 of the Spanish Civil Code). That sentence obviously can only be interpreted as the stipulation of a certain state of affairs (a person's death) as a condition for a certain normative change (the transfer of rights of inheritance). We propose to call such rules — whose canonical form would be 'If state of affairs X obtains, then institutional result (or normative change) R is produced' — purely constitutive rules. And finally, as Josep Aguiló (1990) has clearly shown, sentences like those stipulating the conditions for validly laying down a will, or for issuing a law, do not seem to be restricted to elucidating the meaning in which the legislator uses the terms 'will' or 'law'. The sentence saying that two witnesses are required for a will to be valid does not claim — or does not only, nor mainly, claim — to clarify the sense in which the legislator uses the word 'will', but rather to indicate how someone wishing to attain a certain institutional result must proceed. And something similar could be said about the sentence stipulating the conditions required for issuing a law. If Alchourrón/Bulygin were right, that would mean that sentences like these have no other function than that of identifying certain texts as being 'wills', and others as being 'laws'.[12] But this really sounds very strange, precisely because it does not explain how 'legatees' or 'legislators' use those sentences. We will come back to this in the next paragraph.

In our view, a pragmatically adequate reconstruction of the legal order should distinguish at least between regulative rules[13] (and here, in turn, between principles and rules), power-conferring rules, purely constitutive rules, and definitions.[14] The canonical form of definitions is, in fact, the one suggested by Alchourrón and Bulygin. But that means — we think — that definitions relate words to words (or, if you prefer, to concepts), not cases (states of affairs) to solutions (deontically modalized actions) or conditions (mere states of affairs, in the case of purely constitutive rules, and states of affairs plus institutional actions, in the case of power-conferring rules) to the production of normative results (new states of affairs).

3.2.5. This also enables us to account for legal sentences like 'Article x of law y is herewith derogated', 'Organ X is herewith established', 'I herewith appoint so-

---

[12] And even then there would still be the important difference from definitions, like that of 'being of age', that they do not allow us to identify *norms*, if by norms we understand the *meaning* of norm-sentences, but only texts: that this text is a law, and that one a will, etc.

[13] As the reader will have noticed, we use the term 'deontic or regulative norms' in order to refer to all norms in the consequent of which appears a deontic operator (that is, 'obligatory', 'prohibited', or 'permitted'). In ch. I, we have treated a subclass of deontic or regulative norms: mandatory norms, where this is understood to mean norms containing the operators 'obligatory' or 'prohibited'. The problems concerning 'permissive norms' are analyzed in ch. III.

[14] The distinction between definitions, purely constitutive rules, power-conferring rules and regulative norms has certain similarities with that of Gil Robles (1984) between ontic rules, technical rules, and deontic rules or norms. According to Robles, *ontic rules* are rules stipulating the necessary elements of the respective domain of action (for instance, those stipulating the geographical and temporal limits of the validity of a norm, norms of organization or of competence, etc.); *technical or procedural rules* are rules stipulating the necessary requirements for successfully performing actions within the respective domain; and, finally, *deontic rules* are rules stipulating certain conducts as being obligatory. Independently of the fact that our conception of legal theory has very little in common with Robles' 'extreme formalism' (p. 16), and that the category of deontic rules can be seen as uncontroversial, his typology is different from ours for at least two reasons: on the one hand, he thinks that "there are not really any precepts that are definitions", that is, definitions are not located on the level of legislative language, but on that of the interpreter (p. 223); and on the other, according to Robles, what we call 'power-conferring rules' are a combination of ontic rules (stipulating the static prerequisites — space, time, subjects and competences — of an action) and technical-conventional or procedural rules (stipulating the dynamic prerequisites — that is, the procedure — of an action). A detailed and, in our view, illuminating commentary to Robles' book can be found in Rodilla (1986). See also the reply by Robles (1986).

and-so to such-and-such office', etc. Such sentences— in contexts mostly regulated by regulative norms — express the use of power-conferring rules, and in some cases of definitions, and the production of the corresponding institutional result or normative change. In other words, the issuing of such sentences by the addressees of the corresponding power-conferring rules is, in each case, a *normative act*. These sentences differ from sentences expressing any of the other kinds of norms we have distinguished, because of their performative character: by saying 'I herewith sentence so-and-so to such-and-such punishment', the judge is performing the action of sentencing; by saying 'Article x of law y is herewith derogated', the legislator performs the action of derogation, etc. Note especially that the rules conferring the power to issue norms in a certain area cannot be distinguished from the rules conferring the power to derogate norms in that same area. The derogation of a norm, thus, can be brought about either as the result of an act of derogation (where this means that a so-called particular derogating clause such as in 'Art. X of law Y is herewith derogated' is expressed) or as the result of an act of issuing a new norm which is incompatible with the former, in combination with the application of the criterion of *lex posterior*. So-called generic derogating clauses (as in 'All provisions contradicting the dispositions of the present law are herewith derogated') are pragmatically empty because they do not have any effect that has not already been produced as a consequence of the fact that the new, incompatible law has been issued.[15]

3.3. A functional approach: Power-conferring rules as reasons for action

3.3.1. The distinction we have just drawn between definitions and norms, on the one hand, and between regulative norms, purely constitutive norms and power-conferring rules, on the other, will be clarified — or confirmed — if we move from the structural to the functional level and look at the role any of these entities plays in practical-legal reasoning.

---

[15] On the problems of derogation, see Aguiló (1993). As is well-known, after the 2nd ed. of the *Pure Theory of Law*, Kelsen was forced to add a derogating norm to his typology of norms because derogating had become a 'specific normative function', since a derogating norm — in contrast to other norms — does not refer to a behaviour, but only cancels the 'ought-to-be' of a behaviour stipulated in some other norm. Therefore, a derogating norm looses its validity in the very moment in which it fulfils its function, that is, when the norm it refers to has lost its validity (cf. Kelsen 1973). Josep Aguiló has shown convincingly how all these characteristics of derogation can be understood better if derogation is seen not as a norm, but as a case of a *normative act*.

Obviously, the primary function of norms is that of motivating or guiding the behaviour of people. Undoubtedly, that function is fulfilled by mandatory norms: the (action) norm punishing murder claims, with respect to people in general, to deter that type of behaviour and, with respect to judges, to indicate what they should do in a case of murder. Here, behaviour is, so to speak, guided directly, i. e., by specifying what is prohibited, or what is obligatory under certain circumstances (given certain states of affairs) and sometimes by ordering sanctions for contrary behaviour which then work as auxiliary reasons for reluctant addressees (that is, for Holmes' *bad man* who does not accept norms as guides of behaviour).

In the case of power-conferring rules, the motivation for a certain behaviour works in an indirect (or, as Raz says, indeterminate) way:[16] they do not say directly how we should behave under certain circumstances, but how we can obtain some normative result X; the norm conferring the power to get married, for example, shows the steps to be taken in order to bring about a normative result that can be seen as facilitating a partnership through certain guarantees of stability, economic security, etc. Like technical rules, power-conferring rules are doubly conditional (Conte 1985a, 357; 1985b, 184; Azzoni 1988, 123): they say how we should behave *if* certain conditions obtain and *if* we want to bring about a certain result. Power-conferring rules do, of course, presuppose regulative norms (they would have no functional sense if through them it would not become possible to introduce, derogate, apply, etc., regulative norms), whereas regulative norms make sense by themselves; but that obviously does not mean that one can say that power-conferring rules do not have the function of guiding or orienting behaviour. As Hart[17] had seen very clearly,

---

[16] "The law guides the action of the power-holder himself. It guides his decision whether or not to exercise the power [...] It is because of this fact that power-conferring rules are norms. They guide behaviour. But unlike duty-imposing rules they provide indeterminate guidance. Duties are requirements that defeat the agent's other reasons for action. The guidance provided by powers *depends* on the agent's other reasons. If he has reason for securing the result the power enables him to achieve then he has reason to exercise it. If he has reason to avoid the result then he has reason not to excercize the power [...] Both duties and powers are intended to determine (in different ways) the reasons for or against the actions they affect." (Raz 1980, Postscript, 228 f.; see also Raz 1990, 104 ff.)

[17] "Rules conferring private powers must, if they are to be understood, be looked at from the point of view of those who exercise them. They appear then as an additional element introduced by the law into social life over and above that of coercive control. This is so because possession of these legal powers makes of the private citizen, who, if there were no such rules, would be a mere duty-bearer, a private legislator. [...] Such power-conferring rules are thought of, spoken of, and used in social life differently from rules which impose duties, and they are valued for

they just do this in another way; and that is the reason why they cannot be reduced to those other, directly regulative norms,.

3.3.2. Now, while regulative norms and power-conferring rules are — albeit different types of — guides for action, the same cannot be said of definitions (and, as we will see shortly, with respect to purely constitutive rules, whether or not they are guides for action depends on the nature of the state of affairs operating as the antecedent of the normative change). Again we must turn to the paper by Alchourrón and Bulygin already quoted so often, and again we must say that where they err is in assuming that the *only* function of power-conferring rules is to identify norm-formulations, and that they do not also have a practical function. We can say that Alchourron/Bulygin's error originates in the fact that they see the law exclusively from the point of view of someone reconstructing it theoretically — and who is, therefore, interested in determining what sentences belong to a particular legal system — and not from that of its addressees, such as an ordinary citizen who can be a 'contracting party', a 'testator', etc., a member of parliament who may successfully promote a bill, or a judge who can change the normative situation of certain individuals. They all regard the law as a mechanism that enables them to bring about normative changes; and in our view, one of the criteria for evaluating the (epistemic) quality of a legal theory is — as Hart has pointed out and we already mentioned earlier — whether it is able to account for such points of view. Alchourrón/Bulygin's reductionist attempt looks rather poor when seen from the prism we are looking through now, that is, when norms are regarded as reasons for action. Because while the rule conferring the power to get married does offer a reason for action, i. e., for taking the course of action resulting in 'being married', it is difficult to think that the definition of 'insidiousness' or 'major prison term' are reasons for acting in some way. They are simply mechanisms for understanding the meaning of a norm; thus, they have not a practical, but an explicatory function. And with respect to purely constitutive rules which correlate the production of normative changes with states of affairs — not with actions —, whether or not they are a reason for action depends on whether or not the production of the corresponding state of affairs is under the control of

---

different reasons. [...] The reduction of rules conferring and defining legislative and judicial powers to statements of the conditions under which duties arise has, in the public sphere, a similar obscuring view. Those who exercise these power to make authoritative enactments and orders use these rules in a form of purposive activity utterly different from performance of a duty or submission to coercive control." (Hart 1994, 41)

the agent. For instance, that one comes legally of age at age 18 is no reason for action at all, because the passing of time is beyond the control of any agent. But in other cases, purely constitutive rules can provide reasons for action. That is the case, for example, with the rule that if someone has found a lost object, the owner has the legal obligation of paying 10% of its value to the person who found it (arts. 615 and 616 of the Spanish Civil Code). That rule is an auxiliary reason for looking for lost objects if one wishes to get money.

To sum up the essence of our view of mandatory norms, power-conferring rules, purely constitutive rules and definitions from the point of view of reasons for action, we can say the following. The first — mandatory norms — operate in practical reasoning as categorical imperatives, because for someone who accepts such a norm (for instance, the norm ordering the judge to impose such-and-such a punishment in the case of murder) that is sufficient reason for acting according to it (if the conditions for the application of the norm obtain). Power-conferring rules (and also purely constitutive rules), on the contrary, give rise only to hypothetical imperatives: they are reasons for action if their subject wishes to reach a certain end (a certain normative result). And definitions are no reasons for action at all, but criteria enabling us to understand (or identify) norms (reasons for action).

3.3.3. These distinctions can be shown in a more concrete and — it seems to us — more convincing way if we consider the following two patterns of practical reasoning:

E_1) a) If state of affairs X obtains, then it is obligatory for Z to do Y.
        b) In case C, X obtains.
        c) Hence, in case C, Z should do Y.

E_2) a) If state of affairs X obtains, and if and only if Z does Y, then institutional result R is produced.
        b) Z wishes to reach result R.
        c) Hence, since state of affairs X obtains, Z should do Y.

The following two arguments could be a possible interpretation of the two patterns:

E_1) a) Judges should punish those who have committed murder with a major prison term.[18]
b) The attack by ETA is a case of murder.
c) Hence, judges should punish those who committed the attack with a major prison term.

E_2) a) Persons of age can get married by taking a certain courses of action and only by taking this course.
b) A and B want to get married.
c) Hence, A and B should take that course of action.

Let us now look at these cases somewhat more closely.

i) While in E_1) and $E_{1'}$), it is premise *a)*, that is, the norm (the regulative mandatory norm) that functions as an operative reason, in E_2) and $E_{2'}$), the operative reason is premise *b)*, that is, not the premise stating the norm (the power-conferring rule), but the one attributing certain ends to certain subjects. In E_2) and $E_{2'}$), the normative premise (expressing a power-conferring rule) is simply an auxiliary reason.

ii) E_1), $E_{1'}$), E_2) and $E_{2'}$) are complete arguments, i. e., acceptance of the respective premises necessarily leads to the acceptance of the conclusion, without any need for further premises; in each case, there is an operative reason and an auxiliary reason which, together, constitute a complete reason. However, one thing is that the argument is complete, and quite another that is is closed. With respect to the premise constituting the operative reason, we can ask about the reason for accepting it (if it is a regulative norm, as in the case of premise *a)* in arguments E_1 and $E_{1'}$) or for considering it desirable (if it is the expression of the agent's ends, as in the case of premise *b)* in arguments E_2 and $E_{2'}$); and with respect to any of the premises we can ask what exactly the whole of it or certain expressions employed in it mean (for example, what exactly does 'of age' or 'murder' mean?).

Now, to answer the first question necessarily implies embarking on a new argument with new auxiliary and operative reasons (e. g., art. 405 of the Spanish Criminal Code punishes murder with a major prison; judges ought to

---

[18] [*Translator's note:* A 'major prison term', in Spanish law, is a sentence between 20 years and one day and 30 years. Cf. also the definition of 'minor prison term' in n. 6. According to the Spanish Criminal Code, manslaughter is punished with a minor prison term, whereas murder requires a major prison term.]

obey the valid legal norms; being married to *B* will give *A* economic security; *A* and *B* want *A* to have economic security); the second question, in contrast, can be answered simply with a clarification (e. g., 'of age' means being at least 18 years old; 'murder' is the killing of a person under certain aggravating circumstances, e. g. with malice aforethought etc.); in the second case, we do not obtain reasons for accepting the premises, but — if you wish — for understanding them; what we want to decide with our answer is not whether we should accept the premise in question; rather, we are interested in something prior to that, i. e., in the sense or meaning of the premise. Of course, the question about 'of age' could be posed in another way; we could ask why those who are at least 18 years old are of age, to which one would have to answer that they are of age because some provision thus stipulates it, and that this provision is binding for legal operators. But in that case, what we have done is precisely to modify the nature of the question: we then do not try to *understand* a sentence, but to *justify* why we understand it as we do. Thus, we have gone from the explicatory level (that of definitions as such) to the practical-justificatory level (that of the norms whose meaning is stipulated by the corresponding definitions).

iii) E_1) and E_2) are different schemes of practical reasoning. The fundamental difference between them is that the first, but not the second one, is a deontic pattern. That means that in E_1), premise *a)* is a norm — not a norm-proposition — making it obligatory, prohibited or permitted to perform a certain action, and the same can be said of conclusion *c)*. This is not the case in E_2): E_2) *a)* is not a deontic sentence stipulating that some conduct is obligatory, permitted or prohibited. In a statute, a sentence like 'Persons of age *may* get married' actually is ambiguous. On the one hand, it expresses a power-conferring rule in which the word 'may' does not have a deontic character but only indicates the capacity of changing normative states of affairs; on the other, the sentence also expresses that the exercise of that power is facultative, that is, that the use of it is regulated by a (regulative) deontic norm modalizing the action in that way.[19] One could say that premise *a)* is an abbreviated reference

---

[19] Let us now understand 'Persons of age may get married' as a sentence about the law, rather than a sentence contained in a legal statute: 'According to Spanish law, persons of age may get married'. In that case, it is still ambiguous in the sense that it can mean the following two things:
    a) The law allows persons of age to get married.
    b) The law confers the capacity to bring about the result, or normative change, we express in the phrase 'to be married' on persons of age.
Obviously, the sentence is true if it is interpreted as a norm-proposition referring to a permissive norm as well as if it is interpreted as a norm-proposition referred to a power-conferring rule. But

to an entire set of norms (those concerning the institution of marriage) many of which are of a regulative kind (e. g., the one stipulating the mutual obligation of spouses to support each other). But to this we can answer that those regulative norms are not used but only — implicitly — mentioned in the argument, because one does not need to accept them for the argument to be true. In other words, pattern E_1) is valid only if a regulative norm is used (in the sense that it is accepted); in E_2), regulative norms are also involved, but the argument is valid irrespective of our attitude towards those norms.

The conclusion of E_2) — sentence E_2) c) — is not of a deontic kind either: the word 'should' in this case has only a technical meaning,[20] as when one says that if you want the water to boil then you *should* heat it to 100 degrees. Thus, for example, A and B can come to the conclusion that they *should* take the stipulated course of action because they want to get married in order to provide a legal protection for A who has no savings, no stable employment, etc. However, if A's situation suddenly improves, or if a befriended lawyer suggests another way of providing the desired type of economic security, then it may well be possible that they no longer should take the course of action resulting in marriage; that is, whether or not they should do so does not depend on what regulative norms (or power-conferring rule) stipulate, but on the ends the agents set for themselves.

iv) The agents' ends, however, can be affected by mandatory norms (rules or principles). That is, premise *b)* of pattern E_2) — that Z wishes to attain result R — is something the legal order can make depend exclusively on the

---

the two interpretations are not equivalent; this we can see if we substitute the above sentence with this other one: 'According to Spanish law, the Constitutional Court can declare unconstitutional laws that are constitutional.' That sentence can also mean two different things:

a) Spanish law permits the CC to declare unconstitutional laws that are constitutional.

b) The Spanish Constitution confers on the CC the capacity to annul laws that are constitutional, by declaring them unconstitutional.

Understood as a norm-proposition referring to a permissive norm, the sentence is false, since according to the statutes of the Constitutional Court the CC "is subject [...] to the Constitution" (art. 1.1); however, understood as a norm-proposition referring to a power-conferring rule, the sentence is true, since a declaration of unconstitutionality unfolds its normative effects even if it is itself unconstitutional: "A judicial decision on a complaint of unconstitutionality shall be considered *res judicata*, binding for all public powers, and shall have general effects from the date of its publication in the Official State Gazette" (art. 38.1 of the statute of the Spanish CC).

[20] Some authors follow the convention to reserve the word 'should' for deontic contexts, and to use the expression 'must' for the 'should' we have called technical here. We have not adopted this convention because we think it is useful to reflect the fact that in ordinary language, 'should' and 'must' are used indiscriminately, both in deontic and in technical contexts.

individuals will, or can stipulate as obligatory (as in the case of judges for whom it is obligatory to produce a result, consisting in a legally founded sentence, in the cases they hear). We would then have the case that a mandatory norm becomes the reason justifying the operative reason of that pattern of reasoning. But that, of course, does not impede that E_1) and E_2) must be seen as two different patterns of practical reasoning which, obviously, can be intertwined in a great number of ways and present different degrees of complexity. Just as the law — considered in a static way — cannot be fully understood if we try to reduce all its sentences to one and the same form, legal practical reasoning cannot be reduced to one single pattern.

3.4. Power-conferring rules, non-normative powers, and interests

In order to examine in what way power-conferring rules are connected with power in the social sense and with interests, we will start from the concept of (social) power formulated in ch. I: 'A has power over B if A has the capacity to affect B's interests.'[21] The fact that an (individual or collective) agent has that capacity may be due to several factors: for instance, A's power may be based on his greater physical strength; or on the fact that A possesses certain scarce resources B also desires; or on the fact that B concedes moral authority to A because A is one of B's elders, or for ideological reasons; or on the fact that there is some norm conferring a normative power on A for changing the normative situation of B. What we are interested in here is that last type of social power — normative power — which, of course, is not totally detached from the other kinds: the fact that there is a norm and that it is effective depends on some of the other kinds of power (or a combination thereof). But we will concentrate on normative power — legal normative power — as such, irrespective of its genesis, preconditions, etc.

The power conferred by the legal rules we are focusing on, thus, is a kind of social power, which means that those norms, if used successfully, enable some agent to affect another's, or his own, interests. That is the case with the exercise of legislative power, the power of reglementation, judicial power, etc. Through the use of power-conferring rules, regulative norms can be promulgated or derogated, rights and obligations can be ascribed to certain people, etc. But what must be emphasized is that a regulative norm directly affects the interests of an individual or group (for example, the norm prohibiting homicide

---

[21] See ch. I for further specifications.

secures our interest of not being killed and sets limits to what others may do in order to pursue their interests), whereas a rule conferring a power (for example, that of getting married) gives its addressee the capacity to interfere in the interests of others, as well as his own; hence, the connexion with interests in the first case is a direct one (the mechanism of regulation operates every time a certain state of affairs — the norm's conditions of application — obtains), whereas in the second case, it is indirect (the mechanism of regulation only operates when, in addition to the respective state of affairs obtaining, the power-holder acts in some way). All this, however, is too abstract and needs to be made more specific in several ways.

3.4.1. Like regulative norms (and power-conferring rules), definitions — legislative definitions — are the result of the exercise of a normative power; but seen by themselves, definitions are connected neither with power nor with people's interests in general; they simply allow one to identify (genuine) norms or to clarify in what way they do it. What affects people's interests is not that 'major prison term' is understood to mean a term between 20 years and one day and 30 years, but the existence of norms punishing certain acts with a major prison term. Similarly, the mere definition of 'being of age' as being at least 18 years old does not affect in the least bit the powers people may have; what is relevant in this case is that when turning 18, a person undergoes the normative change of coming 'of age', that is, she becomes the subject of rules conferring on her the power to manage her own possessions, to make contracts, to vote and be eligible for public office, and so on. In fact, with respect to definitions (and speaking of people in general), the element of power is found either before (in the rule conferring the power to stipulate them) or after (in the regulative rules that stipulate the criteria for how to interpret legal sentences, including definitions), or in the (regulative, power-conferring or purely constitutive) norms whose meaning definitions help to uncover or stipulate themselves.

Nevertheless, definitions as such do affect the conformation of the respective power of law-creating organs, on the one hand, and of those who interpret and apply the law, on the other, because by precisely establishing the meaning of the terms used in legislative language, definitions fulfil the function of reducing the 'semantic power' of judges and legal doctrine. As Mario Jori has written, "the modern dispute about legislative definitions certainly leads to a political struggle over how much power the legislator has, or ought to have, over the words of the law, and therefore, over the limits of his control of the law [...] Semiotic choices, the choices about whether, who and how to define legislative terms, in the last instance, imply ethical-political evaluations" (Jori 1995, 143 f.).[22]

3.4.2. The distinction between private power and public power can obviously be drawn according to a number of different criteria (for example, according to who exercises power, how it is exercised, for what ends, etc.). We will not go into this difficult question here; but we do wish to point out that our definition of power covers both the case of what we can call *power of heteronomy* — that is, when A and B are different subjects — and *power of autonomy*, that is, the power of self-determination where A and B are one and the same agent. Public power is basically power of heteronomy, and private power is power of autonomy, but this is not always so: the power of going to court — power of heteronomy — does not seem to be a clear-cut case of public power, and neither does parental power; and the power of a public organ to organize itself in some way — for example, to give itself a statute — does not seem to be a case of private power. In any case — to repeat it once more — we are not so much interested in the question of how to distinguish between the two types of power, as to show that the (extremely wide) notion of power we are using enables us to cover all types of cases in which one usually speaks of power-conferring rules.

3.4.3. The distinction (among regulative norms) between permissive norms and mandatory norms (i. e., those stipulating duties and prohibitions) is of special interest here since, as we have pointed out at the beginning of the chapter, several authors have seen power-conferring rules as permissions. In our opinion, the distinction between these two types of (regulative) norms and power-

---

[22] According to Jori, modern law is the most important case (other examples are "the theological discourse of dogmatic religions, political discourse controlled by party organizations, and even some natural languages insofar as one could think that they are administered by an academy of purists"; Jori 1995, 131) of what he calls 'administered languages', that is, languages belonging to an intermediate category between instrumental-artificial languages and natural languages; languages belonging to that intermediate category are characterized by their need for authorities that 'administer' them (in the case of modern law, such authorities are the legislator, the judges, and doctrine), in contrast to what happens with technical languages, or with natural languages which administer themselves. Jori thinks that only by taking into account the pragmatic aspect of language (the category of 'administered language' is a pragmatic category) which has traditionally been ignored in legal culture, we can elaborate an adequate legal semiotics and thus a satisfactory analysis of definitions. Considerations similar to those of Jori, on definitions as instruments for constraining the power of law-applying organs can be found in Guastini (1985).

conferring norms can be drawn as follows: Mandatory norms, by stipulating duties or prohibitions, impose restrictions on their addressees' ways of pursuing their — individual or collective — interests, that is, they directly constrain the physical, economic, etc. power of their addressees; permissive norms, in contrast, impose prohibitions of interference on others: they thus guarantee that every agent can pursue his — individual or collective — interests without interference from others; and power-conferring rules enable agents, within certain limits, to modify their own or others' normative position and thus influence his own or others' interests.

One could then think that the first two (those directly or indirectly stipulating prohibitions and obligations) imply a restriction for the persecution of interests (for those subjects that ought to perform the obligatory action, or abstain from the actions that are prohibited), whereas power-conferring rules have the opposite function, that is, in some way they help their addressees pursue their interests. But that idea is actually wrong. The rule conferring on parents legal power over their children obviously does not have the function of helping parents pursue their own interests, but those of their children. And something similar can be said of the power of judges to sentence, of parliament to legislate, or of the administration to issue regulations. This confusion, as we already said, probably has to do with the ambiguity of the expression 'can', which can mean either 'permission' or 'capacity' to obtain certain results in the natural or in the normative world. The 'can' that appears in a permissive norm is that of doing, or not doing, certain actions without being hindered by others; the opposite to this sense of 'can' would be that the action in question is obligatory or that it is prohibited. This is why, in contrast to mandatory norms, the addressee of a permissive norm cannot fail to comply with it. On the other hand, the 'can' of a power-conferring rule is that of attaining certain normative results by performing a certain action, under certain circumstances, which may in turn be permitted, obligatory, or prohibited; the opposite to 'can' in this second sense is to be incompetent, that is, not to have the capacity to bring about a certain normative result. Finally, it is also impossible not to comply with a power-conferring rule, but not for the same reason as with permissions, but simply because they are not deontic norms: the only thing one can do with a power-conferring rule is to use it successfully, or not to use it.

APPENDIX TO CHAPTER II
REPLY TO OUR CRITICS

1. Introduction

An earlier version of our ideas about power-conferring rules (Atienza/Ruiz Manero 1994a) has been critically assessed by several of our colleagues connected with Pompeu Fabra University of Barcelona: by Ricardo Caracciolo (1995) on the one hand, and Daniel Mendonça, Juan Moreso and Pablo Navarro (1995), on the other. Since it is a privilege to have such strong theoretical adversaries, in the following pages we will do our very best to deepen even more this relationship of opposition, which, in any case, will never become as strong as our friendship with all of them on the personal level. There may be some truth in the saying that only in adversity you will really know your friends.

The main part of the contribution by Mendonça/Moreso/Navarro reproduces the structure of our paper — which has now become ch. II of this book — and thus is divided into three parts: In the first two, they criticize the *pars destruens* of our article, that is, the way we present what we consider the to be the explicatory deficits of two alternative conceptions of power-conferring rules: that which we referred to as the deontic (and which they prefer to call the 'prescriptivist') conception, and the conceptualist conception. In the third part, they criticize our proposal to understand power-conferring rules as anankastic-constitutive rules. In our reply, we will follow that same order, and we will try at the same time to reply to the observations made by Ricardo Caracciolo.

2. Critique of our critique of the deontic (or prescriptivist) conception

2.1. Our central criticism against the deontic conception of power-conferring rules (that understands them as permissive norms) was that such a conception does not enable us to account for the difference between an *irregular use* of a normative power (the case of a judicial decision *contra legem* issued by a court on a case falling under its jurisdiction, or of an unconstitutional law enacted by a parliament) and a *non-use* of any normative power whatsoever (the case of the 'judicial decision' issued in a class of practical exercises by a student, the 'law' approved by an assembly of professors, the 'declaration of war' against some foreign power issued by the judge of a criminal court). Awareness of this difference is, in our view, deeply rooted in the conceptual intuitions of everyone dealing with law. In addition, it seems that these conceptual intuitions hint at an important characteristic of reality, namely, that as long as there is no use of any normative power the normative universe remains unchanged (when a judge declares war on Serbia, for instance, this may give rise to a scandal and to questions about the judge's mental health in the press, but the normative universe regulating the relations between Serbia and Spain remains unchanged), whereas in the case of an irregular use of a normative power, a normative change *is* produced (if a criminal court after hearing a murder case issues a sentence *contra legem*, this does change the normative situation of the accused, irrespective of whether or not the decision in question can be appealed). Following von Wright, Mendonça/Moreso/Navarro say that it is "plausible to understand theoretical reconstructions as an explication of our conceptual intuitions".[1] But in that case, it is obvious that a theoretical reconstruction of power-conferring rules that is shown to be unable to account for that difference — as is the case with the conception seeing them in terms of normative permissions — is a theoretical reconstruction which, for that reason alone, should be abandoned. Now, Mendonça/Moreso/Navarro nowhere try to show that the deontic (or, in their terminology, prescriptivist) conception can account for that — in our view, crucial — difference. Instead they argue that, since the deontic conception does not claim to account for it, it cannot be required to do so. In their own words:

"For the prescriptivist conception, the existence of legal effects is neither necessary nor sufficient for attributing competence to a norm-authority. But if there is no conceptual relation between the two phenomena, then it cannot be right to say that the prescriptivist conception of norms of competence is deficient because it does not account for those legal effects."

But then, we admit that we fail to see what a theory of power-conferring rules (or rules of competence) could account for, if not the capacity to produce normative changes.

---

[1] Their acceptance of von Wright's meta-theoretical criterion, however, does not at all seem to fit to the assertion — apparently intended to dismiss the value of widely shared conceptual intuitions — that "whether all, some or no-one accepts a decision *contra legem* or a declaration of war by a court is an empirical question the truth of which, therefore, is contingent" (p. 8). Of course, questions such as whether there is such a thing as a planet Earth, whether people are living on it, whether these people form societies, whether a certain society has a legal system, whether that legal system contains norms of competenece permitting deliberate changes of its norms, and other questions of this kind are also empirical questions, and the truth of assertions about them, obviously, also is contingent.

2.2. When referring to different types of irregularities of norms, Mendonça/Moreso/Navarro perhaps inadvertently use the distinction between power-conferring rules (or rules of competence) and deontic or regulative rules referring to the exercise of such a competence — a distinction which in the context of a conception of power-conferring rules as deontic rules makes no sense. They say that "the expression 'irregular norm' can be associated with different states of affairs: *i)* a prohibited action performed by an incompetent subject; *ii)* an action that is not prohibited but is performed by an incompetent subject; *iii)* a prohibited action performed by a competent subject". Now, on the deontic or prescriptivist conception, the set of normative acts a subject is competent to perform is identified with the set of normative acts that subject is permitted to perform. So, from this perspective, sentence *i)* would be merely a tautology, while sentences *ii)* and *iii)* would be contradictory.

2.3. Mendonça/Moreso/Navarro ask whether our thesis is "(T_1) In order to attribute competence to the organ that has issued an irregular norm N, it is necessary that N produces legal effects" or rather "(T_2) In order to attribute competence to the organ that has issued an irregular norm N, it is sufficient that N produces legal effects". They seem to assume that by 'legal effects' we understand 'appealing the validity (legality) of a normative act', and on several occasions (e. g., in n. 3) they complain that we have not made it clear what we understand by 'legal effects'. It may be true that we have not been very clear on that point; but we think that anyone concluding that by 'legal effects' we understand *the fact* of filing an appeal must have read our paper in great haste. We also do not identify 'legal effects' with *the possibility* of filing an appeal, for the obvious reason that this possibility does not exist, for example, against a decision issued by a court of last instance, and no-one would doubt that issuing such a decision is a clear case of a normative act. By 'legal effects' we understand the effects the law, in each case, correlates with the performance of a certain normative act (issuing a judicial decision, presenting a bill of law, enacting a law, making a will ...). And with respect to the possibility of appeal, what we hold is that the existence of that possibility presupposes either the performance of the corresponding normative act (what is disputed in that case is the regularity of its result) or some kind of appearance of a performance of such a normative act (what is disputed then is that such a performance has taken place, which implies disputing the very existence of the institutional result, irrespective of its regularity or irregularity). Once this is clear, it is, in our view, a necessary and a sufficient condition for attributing competence to an organ that has issued a (regular or irregular) norm that that norm produce legal effects, in the sense just explained — provided, of course, that we are talking about a norm that exists as the result of a process of enactment (that is, not a last norm, or a norm of common law, or a norm that exists merely as a logical consequence of other norms, since in those cases we cannot speak of enactment).

The case of the will Mendonça/Moreso/Navarro refer to can be used as an argument against what has just been said — and maybe they see it in that way. The case, as presented by our critics, is as follows. Suppose there is a norm stipulating that a necessary condition for performing the normative act of 'making a will' is the presence of two witnesses. Thus, our authors write,

"if the making of one's last will does not satisfy the requirement of the presence of two witnesses, then it is not a 'will'. However, as a question of fact, such a disposition can have the same effects as a will. Suppose a declaration of one's last will that is contested in court because it has not been put down in the presence of two witnesses, is declared valid by the judge. A-RM should say that the judge's decision constitutes an erroneous application of the law: a sentence *contra legem*, issued by a competent authority. But with respect to the disposition of last will, they adopt the opposite solution: even if that declaration is qualified as a 'valid will' by the judge and has full effect, A-RM would not accept that that declaration of last will is a will."

Our critics reconstruct correctly what would be our position on that case. Because we certainly do not believe — and here we also reply to an observation made several times in Caracciolo's paper — that judicial declarations are a criterion of truth; but in our view, that does not pose any special problem for our conception. Since, in fact, if the normative act of 'making a will' has as a necessary condition that it must be done in the presence of two witnesses, then obviously in the case in question a will has not been made, no matter what the judge says. If the conditions stipulated as necessary for the performance of some normative act by a power-conferring rule are not satisfied, then the assertion that they have been satisfied and that the normative act has been performed is false, irrespective of who pronounces it. What happens is that, because of the normative powers of the judge, that circumstance becomes legally irrelevant (leaving aside for a moment the possibility of appeal). But that is just the same as in the case of a judge's statements about whether or not some regulative rule has been violated. If John has not killed Peter, then the sentence 'John has killed Peter' is false, even when it is pronounced by a judge in a justification for a decision. But that does not mean that John cannot be convicted of murder and that (except for the possibility of an appeal) the fact that the decision is wrong cannot become irrelevant for John's legal situation. It is the duty of the judge to qualify as a 'will' those, and only those, declarations of last will that satisfy certain conditions, and to convict of murder those, and only those, who have committed murder. But when — by mistake or intentionally — he violates these duties, his decisions are still judicial decisions, that is, they produce the legal effects the law has connected with their being issued. What the example shows is that — even in the case of the normative power to determine whether someone else has made use of another normative power — to *use a normative power* and to *use it regularly* are not the same.

2.4. Since we wanted to stress the difference between rules stipulating necessary, sufficient, or necessary and sufficient conditions for the performance of some normative act (that is, for bringing about some institutional result or normative change) and regulative rules discriminating between licit and illicit institutional results, we probably did not sufficiently elaborate the possibility that an appeal may contest the very existence of a normative result (that is, that the conditions a power-conferring rule stipulates as sufficient, or as necessary and sufficient, for the corresponding normative act have been satisfied), rather than its regularity. This seems to be the cause of some of Caracciolo's understandable perplexities. He quotes a passage where we say that "the very fact that unconstitutional laws or illegal judicial decisions can be appealed shows that such laws or judicial decisions are precisely what they claim to be, that is, laws or judicial decisions. Because there is no appeal against a 'law' issued, let's say, by a group of professors, or a 'judicial decision' issued in an exercise class in criminal law by an undergraduate student". He then comments that "this passage implies the implausible consequence that invalid acts, in the sense of V_1 (those that do not satisfy the rule of competence), cannot be appealed". We do not think that that consequence is implied by our statements. What we want to say is that in those cases where it is clear that the normative act has not been performed (because, for instance, the subjects were not qualified), obviously, no appeal is possible. But power-conferring rules, as any other norms, can, of course, be open-textured and thus have a zone of penumbra which can make it contestable whether or not the normative act has been performed; and therefore, legal systems foresee the possibility of an appeal in that case too. It should be noted, though, that in such a case, what is disputed is not whether or not the institutional result is irregular, but something prior, i. e., whether it exists at all.

2.5. We think that, generally, our critics' use of deontic operators in the context of power-conferring rules is flawed because they do not distinguish between the *action* element and the *result* element. We will try to explain this, starting with what we have called the 'canonical form' of a power-conferring rule: 'If state of affairs X obtains, and Z performs action Y, then institutional result (or normative change) R is produced'. Now, consider element Y of this formula, that is, the action the power-conferring rule links to the production of the respective normative result. To speak of an 'action' in this context obviously is a simplification. So we should be a little bit more precise: Power-conferring rules usually do not link the production of the normative result to one single action, but rather to a certain conjunction of actions (a course of action), to a disjunction of actions or courses of action, or to some combination of both.[2] In this context (that is, with respect to the relationship between the Y element and the R element of a power-conferring rule), to say that some course of action (some conjunction of actions) or some part of it is obligatory means that that course of action or that part of it must necessarily be followed in order to bring

---

[2] Naturally, all this can be described as a unitary action leading to the result; the result 'having gotten married', for example, corresponds to the action 'getting married'. But it is obvious that in the power-conferring rule the action cannot be described in this way, since in that case it would say that if someone (who fulfils certain conditions like being unmarried, being at least of a certain age, etc.) gets married, then that someone has gotten married. In the formulation of the power-conferring rule, the action must be described independently of the result, that is, as a natural action, or as an institutional action depending on some other rule (e. g., in our case, to fill in certain papers, promise certain things before a judge, etc.) (We are here taking into account a very pertinent observation made by Ricardo Caracciolo in n. 5 of his paper.)

Daniel González Lagier has suggested another way of expressing this, using von Wright's distinction between *activity* and *action*. On this distinction, von Wright says the following: "As acts are related to *events*, so are activities related to *processes*. Events happen, processes go on. Acts effect the happening of events, activities keep processes going" (von Wright 1963, 41); "Action may be said to presuppose or require activity. The bodily movements which are a prerequisite of most human acts may be regarded as activity in which the agent has to engage in order to perform those acts. The changes and states which we call *results* of action may be viewed as *consequences* of such prerequisite activities." (ibid.) To speak of an activity as a sequence of bodily movements is a simplification (or an extreme case): an activity is all that an agent does in order to be able to carry out an action. The activity produces the result of the action; so the relationship between the activity and the result is always a causal one (or a conventional one, in case of an institutional action); but seen from the perspective of the action, its relationship with the result is a logical or conceptual one (since we use the result in order to identify that action). Using these categories, then, we could say that a power-conferring rule indicates what the necessary (and, perhaps, also sufficient) *activity* for performing the *action* (which is now described by pointing to the result: getting married) is.

about the normative result (that is, that the course of action or that part of it is a necessary — and possibly a sufficient — condition for the result); whereas to say that the norm subject is permitted to choose among several courses of action — or that, within some course of action, in a certain part of it he may choose between different subcourses of action — means that following *some* of those courses or subcourses of action is a necessary — and possibly a sufficient — condition for the result. 'Obligatory', 'permitted', and 'prohibited' are not of a deontic, but of an anankastic character here:[3] in the first case, they indicate that in order to bring about the result, the norm subject *must* follow a certain course of action, and in the second, that in order to bring about the result, it *may* — in a non-deontic sense — choose between different courses or subcourses of action and that it *must* choose one of them; 'prohibited' then indicates that a certain course of action is inadequate for bringing about the result.[4]

In contrast, when referring to the result, 'obligatory', 'permitted', and 'prohibited' are deontic operators — deontic operators contained in rules that presuppose the effective production of the result, that is, that given state of affairs X, subject Z has performed one of the courses of action Y that are a sufficient, or necessary and sufficient, condition for the result.

2.6. We think that this can help us restate the problem of the 'judicial decision' issued by the student and solve a problem pointed out by Caracciolo, concerning the ambiguity of some legal provisions which, he says, can be interpreted either as permissive norms or as power-conferring rules. Let us use an, obviously, extremely simplified model of a power-conferring rule and suppose that X is the fact that someone is standing trial, Z is the judge in charge of the trial, Y stands for writing a document, consisting in 'material facts of the case',

---

[3] It is worth recalling that in some papers von Wright, in a way that seems to be compatible with what we have sustained here, mentions the possibility that in some norms, the concepts 'obligatory', 'prohibited' and 'permitted' are not of a deontic nature. More precisely, in *On the Logic and Ontology of Norms* (von Wright 1969) he refers to "rules for making contracts; the formalities which have to be gone through if a marriage is to be legally valid; the rules concerning the appointment of officials to offices and concerning the professional qualifications which the holders of offices have to satisfy; many, perhaps most, rules of civil and criminal procedure; and, finally, many rules belonging to constitutional law such as, e. g., the rules according to which laws have to be enacted" (ibid., 97). The 'ought' those norms refer to, according to von Wright, can be understood as meaning something like 'unless an individual i does action A or has property P, he does not count as an agent of a certain category' (ibid., 98 ff.).
[4] Of course, not all procedural rules stipulate necessary, sufficient, or necessary and sufficient conditions for the production of a normative result. There are procedural rules whose violation does not imply that the normative result does not exist, but only that it is irregular. Therefore, such rules are not anankastic-constitutive (power-conferring) rules, but only regulative rules.

'legal foundations', and 'ruling', and R means the institutional result of a 'prison sentence' referring to the person tried. In the case of the student, since element Z of the antecedent of the power-conferring rule is missing, the institutional result 'prison sentence' simply cannot be brought about, even if state of affairs X obtains and the student performs action Y. But we cannot say that action Y is prohibited to the student: there is no norm prohibiting to 'play judge' by writing such a document.[5] But if, in contrast, X obtains and the judge is the one actually entrusted with the case, then X plus Y produce the institutional result 'prison sentence'. That institutional result may be the one the judge ought to bring about (it is his obligation to bring about) — that is, a sentence grounded in the law — or it can be a prohibited institutional result (a sentence *contra legem*).

Ricardo Caracciolo poses the problem of provisions which, in his view, are formulated in an ambiguous way that can be interpreted either as a permissive norm or as conforming to our 'canonical form' of a power-conferring rule, and he says that "in the absence of an additional criterion that could resolve the ambiguity, there is no way of saying that one interpretation is better than the other". The example Caracciolo gives is the following:

Under conditions X, the members of class Z are *authorized* to perform action R by way of procedure Y.

Our answer would be that if 'to perform action R by way of procedure Y' means that the underlying action Y counts as the institutional action R (or, to say it in a more intuitive way, that by way of action Y institutional result R is produced) then such a provision can no longer be interpreted as a power-conferring rule, that is, a rule that constitutes the possibility of institutional action R, indicating that Y is a sufficient (or possibly a necessary and sufficient) condition for this. Besides, of course, a power-conferring rule not always needs to be accompanied by regulative rules deontically modalizing its exercise (or the modalities of its exercise). If there are no such regulative rules, the exercise of the power in question simply is free from any normative restrictions. In any case, in our view, the constitution of the possibility of institutional action R is a precondition for any deontic modalization of it. This is nothing but an application of the

---

[5] Therefore, the 'non-tautological answer' offered, according to Caracciolo, by the 'prescriptivist conception' is clearly false: such a conception would say that — in Caracciolo's own words — "only judges are *permitted* to carry out the actions underlying an act which is characterized as the issuing of a judicial decision, that is, that there is a permissive norm of which they are the addressees".

general idea — certainly shared by Caracciolo — that the deontic 'can' presupposes the aletic 'can'.

3. Are we treating the conceptualist thesis fairly?

3.1. Mendonça/Moreso/Navarro say that we "do not distinguish clearly between the properties of conceptual rules and the properties of semantic rules" and that "since [we] do not clearly distinguish between different kinds of conceptual rules, [our] treatment of the matter is not fair with those who defend a conceptualist approach". The only thing we can say to that here is, first, that we have not tried to debate against the best conceivable version of the conceptualist thesis, but against the most reliable versions of it in contemporary legal theory we are aware of. And second, that in the version of Alchourrón/Bulygin, what is stressed are not the differences between the different kinds of conceptual rules, but precisely what they have in common (as is shown, for example, in the passage by Bulygin quoted by them: "what all these [conceptual] rules have in common is their defining character, that is, they can be seen as definitions of certain concepts"). If the objection of Mendonça/Moreso/Navarro is that, instead of improving the conceptualist thesis before criticizing it, we took it just as it is presented by its most representative defenders, they are certainly right. But in that case, we are grateful for the excuse they themselves offer: "the missing distinction in A-RM could be partly the responsibility of Alchourrón and Bulygin, since they do not analyze clearly what the defining or determinating property conceptual rules share consists of".

3.2. Mendonça/Moreso/Navarro think it "rather strange" that we attribute to Alchourrón and Bulygin "the idea that a legal order can be satisfactorily reconstructed with the help of two categories of norms: norms of conduct and conceptual rules". In our critics' opinion, "Alchourrón and Bulygin have insisted that legal systems can contain a substantial variety of sentences". In order to support this claim, they quote a passage from Eugenio Bulygin where he says — obviously speaking in general and not specifically of the law — that "there can be many classes of rules". What happens is that this does in no way invalidate our interpretation. As far as we know, no other category besides rules of conduct and definitions has ever been used by Alchourrón and Bulygin to account for legal sentences — except for the fact that in some of their works they mention "sentences presenting political theories, expressing a people's gratitude to the head of State, or asking for God's protection ..." (Alchourrón/ Bulygin 1974, 107); but such sentences "have no influence whatsoever on the normative consequences of the system" and therefore, they can be considered irrelevant.

4. Problems with our conception

4.1. Mendonça/Moreso/Navarro correctly qualify the characterization of power-conferring rules as "anankastic-constitutive rules that can be used as technical-institutional rules" as "a point of crucial importance" of our conception. But to that they reply *1)* that the corresponding technical rule does not belong to the legal system since it does not satisfy any criterion of membership, and that our assertion that the power-conferring rule gives way to a technical rule is an assertion about the use of legal rules by individuals and authorities, and *2)* that, if this is so, then "there are no differences between norms of competence and other legal provisions, since any of them can give way to the elaboration of a technical rule". As for point *1)*, they are certainly right, and that is precisely as we presented things in our text: just as an anankastic proposition is the basis for a natural technical rule for anyone wishing to induce some natural change, a power-conferring rule — an anankastic-constitutive rule — is the basis for a technical rule for anyone wishing to induce some normative change. As for point *2)*, we think that the answer to their assertion that any legal rule can give way to the elaboration of a technical rule is already contained explicitly in our text, where we examine the case, introduced by MacCormick, of that fictional character who, knowing that a certain conduct constitutes a crime and wanting to commit a crime because he wants to be sentenced and make himself comfortable in prison for a while, he performs that action. That person (a somewhat peculiar kind of Holmesian *bad man*) extracts technical rules from regulative norms, as does generally that species of the more common *bad man* whose sole purpose it is to make sure that the law favors (or does the least possible harm to) his interests. To that, two things can be replied, as we did in our text: *1)* that the distinction between regulative norms and power-conferring rules can only be drawn if one takes into account the purpose generally attributed to the provision in question. Committing a crime also (among other things) produces a normative change in the situation of the perpetrator. But committing a crime cannot be seen as an instance of the use of a power-conferring rule, because in general it is understood that criminal norms are not issued with the purpose of enabling people to change their normative situation, but with that of deterring them from performing certain types of actions; *2)* that while from the perspective of a *bad man* one may think that any legal norm offers the basis of a technical rule, from the perspective of a *good man* the existence of a regulative mandatory norm implies the existence of a categorical — i. e., a genuinely normative — duty, whereas the existence of a power-conferring rule implies the existence of a mere hypothetical, i. e., a technical duty.

4.2. Mendonça/Moreso/Navarro criticize that we use argument patterns (E_1 and E_2)[6] without first explaining in what sense they can be said to be valid. In particular, they say that E_2 is invalid, for which — if we have correctly understood their critique — they give three reasons. The first reason has two parts: on the one hand, in premise *a)*, the connection between *Y* and *R* should be that of a necessary, not a sufficient condition; on the other, an additional premise *b')*: 'State of affairs *X* is produced' should be added in order to reach conclusion *c)*: '*Z* should do *Y*'. Our critics think that this error, or carelessness, on our part is due to our "conception of norms of competence: For them [= for us], those norms state 'the sufficient (or necessary and sufficient) conditions for attaining the result, and [...] a norm merely containing some of the necessary conditions [is] only an incomplete norm'". The second one of their reasons seems to be derived from our assertion that "in E_2), regulative norms are also involved, but the argument is valid irrespective of our attitude towards those norms". To this they object that "it is important to note that premises *a)* and *b)* of pattern E_2) make no reference to prescriptive (regulative) norms and — *pace* A-RM — do not imply (nor are implied by) prescriptions". Finally, the third, and weightiest, reason is expressed as follows: "A-RM's mistake probably comes from their not distinguishing the possible logical relations between a set of sentences and the

---

[6] Those patterns were as follows:
E_1)   a)   If state of affairs *X* obtains, then it is obligatory for *Z* to do *Y*.
       b)   In case *C*, *X* obtains.
       c)   Hence, in case *C*, *Z* should do *Y*.
E_2)   a)   If state of affairs *X* obtains, and if and only if *Z* does *Y*, then institutional result *R* is produced.
       b)   *Z* wishes to reach result *R*.
       c)   Hence, since state of affairs *X* obtains, *Z* should do *Y*.
The 'if and only if' in premise *a)* of pattern E_2) appeared in the Italian version of our paper (i. e., the one criticized by Mendonça/Moreso/Navarro) only as 'if'; also, in that version, in the conclusions of E_1 and E_2 'in case *C*' and 'since state of affairs *X* obtains', respectively, was missing.

attitudes of individuals with respect to those sentences. The acceptance of a norm is a fact. The premises and conclusions of an inference are propositional entities. The properties of that set (that is, whether they are true or false, accepted or rejected, credible or improbable, etc.) should not be confused with the properties of the inference."

With respect to the first critical observation, it is true that the connection between Y and R cannot be that of a sufficient condition; that was a mistake we corrected ourselves in the Spanish version — as Mendonça/Moreso/Navarro acknowledge in their paper —, so we have nothing to add to that. It is also correct that premise $b')$ must be added in order to complete the argument; but that is an observation that does not seem to be important: in our pattern, that premise was simply contained implicitly, so that is no reason to question the argument's validity. But nothing of this affects our conception of power-conferring rules. Premise $a)$ of pattern $E_2)$ is a technical rule arising from the use of an anankastic-constitutive rule (the rule conferring the corresponding power). And that last one, in fact, is complete only if it determines the sufficient, or necessary and sufficient, conditions for obtaining the institutional result. But an agent may well argue from an incomplete power-conferring rule (stating only necessary conditions for the result) and still do this in a valid way.

The second of their objections seems to be only a misunderstanding. When we wrote that "in $E_2)$, regulative norms are also involved", what we wanted to say — and what we think we did say — was that course of action Y and result R can also be subjected to regulative norms that qualify them deontically (as a whole or in some of their elements), but not, as our critics seem to have understood, that they imply or are implied (in the logical sense) by regulative norms. Actually, we did not use the word 'implication' in the sense of 'logical implication' (and only an essentialist would say that there is no other implication than logical implication, and that we would commit a conceptual — and not only an empirical — mistake if we would say, e. g., that our critics are implied in a crime, say, of defiling graves).

Finally, with respect to their third observation, the mistake our critics make here is probably due to the fact that they link our pattern $E_2)$ to von Wright's 'practical syllogism' which, as they themselves indicate, stipulates that "if an individual considers an action M necessary for an end F and wishes to attain F, then she is compelled to do M". Now, the conclusion of our argument pattern is not a fact, i. e., does not consist in the fact of accepting a norm, but an ought sentence, even though a non-deontic one — as we explained in the text

— since the 'should' of 'Z should do Y' has only a technical meaning. Besides, we too think that the reasoning done by an agent is one thing, and quite another whether, in fact, he feels compelled to do Y. It is also clear, however, that being different things does not mean that they have nothing to do with each other. But that is another matter.

4.3. In our paper we sustained the thesis that legal definitions are not reasons for action of any kind, but merely help to identify norms, which are the only reasons for action the law provides. Mendonça/Moreso/Navarro propose the thesis that definitions can also be auxiliary reasons. They illustrate this with the following example:

a) By 'rural estates', we will understand estates located more than 50 km away from an urban center.
b) X wishes to acquire a rural estate.
c) Z is the only estate located more than 50 km away from an urban center.
d) X should acquire Z.

Now, if 'estate located more than 50 km away from an urban center' is a definition of 'rural estate', then by the definition of 'definition', premise *b)* says precisely that 'X wishes to acquire an estate located more than 50 km away from an urban center'. That means that *a)* only identifies the meaning of *b)*, which is the premise expressing the operative reason in this argument. And this argument has no other auxiliary reason than the one expressed in *c)*, because only this premise fulfils the specific function of premises expressing auxiliary reasons and which — as Joseph Raz asserts — "is to justify, as it were, the transfer of the practical attitude from the statement of the operative reason to the conclusion" (1990, 33); *a)* neither does nor could fulfil this function; only the defining premise, that is, the one stating the meaning in which the terms contained in the other premises are used, can do this.

4.4. At the end of their paper, Mendonça/Moreso/Navarro ask three questions:

"i) The ontological question: Is the competence of a norm-authority NA necessary or sufficient for the existence of a norm N?"

"ii) The systematic question: Is the competence of NA necessary or sufficient for a norm N to belong to a legal system LS?"

"iii) The semantic question: Is the competence of NA necessary or sufficient for qualifying a norm N as a 'law', a 'judicial decision', etc.?"

As for the ontological question, the competence of an authority obviously is not a necessary condition for the existence of norms: there are norms — ultimate norms, customary norms, norms derived logically from other norms — that are not the result of an act of exercising a competence. In the case of norms that exist because they are the result of acts of prescription, however, the competence of the issuing authority is a necessary (and, together with the fact that the state of affairs mentioned in the antecedent of the power-conferring rule obtains and the corresponding action is performed, also a sufficient) condition for their existence.

As for the systematic question, if, following Alchourrón/Bulygin and Caracciolo (cf. Alchourrón/Bulygin 1991 and Caracciolo 1988 and 1991), by 'belonging' we understand either 'direct belonging to the system', in the case of original norms, or 'satisfaction of the criteria of deductibility or legality and non-derogation', in the case of all others, then the fact that a norm has been prescribed by a competent authority is not a necessary condition for belonging, nor is its conjunction with the other conditions mentioned in the antecedent of the power-conferring rule a sufficient condition, because irregular norms satisfy neither the criterion of legality nor that of deductibility.

As for the semantic question, if it is restricted to norms like 'laws' or 'judicial decisions' — that is, norms existing as the result of normative acts — then the competence of the norm-authority, together with the other conditions mentioned in the antecedent of the power-conferring rule, are a necessary and sufficient condition for being able to speak of a 'law', a 'judicial decision', etc.

Chapter III
Permissive Sentences

After having treated, in the first two chapters, the problems of mandatory norms — rules and principles — and of power-conferring rules, purely constitutive rules and definitions, we will now set out to examine permissive sentences. In order to do this, we will first present what, in our view, are the main positions on this matter in contemporary legal theory, and then develop our own conception. Our central point will be that with respect to permissions one must distinguish three contexts: that of rules regulating 'natural' conduct, that of power-conferring rules, and that of legal principles.

1. Permission in contemporary legal theory

The question of whether there is such a thing as permissive norms is one of the most hotly debated issues in legal theory. The question, of course, is not whether in legal systems there exist sentences of the kind 'p is permitted' or 'p is facultative'. Obviously, sentences of that kind are perfectly common in legal systems. The problems giving rise to doubts about the existence of permissive norms (or, if you prefer, about the need for the category of 'permissive norms'), rather, are the following two:

1) As everyone knows, the primary function of any normative systems is usually understood to be that of guiding human conduct. In the case of mandatory norms,[1] that function is fulfilled by stipulating either the obligation to perform a certain action p in a certain case q (or, what amounts to the same, the prohibition to forbear p in q) or the prohibition to perform p in q (or, what is the same, the obligation to forbear p in q). Thus, mandatory norms, which can be expressed in the form of obligations or prohibitions, command either to do some action or to forbear it, and in this way they separate the sphere of what is licit from that of what is illicit. Permissive norms, understood as norms permitting performance as well as forbearance of some action p in some case q, in contrast, do not command anything. Now, does this mean that when we are confronting a norm of this kind — i. e., a norm that permits the performance,

---

[1] As the reader will see presently, the line of argument developed in the text refers to *action norms*, but it also applies to *end norms*, if one replaces all reference to action p by a reference to some state of affairs F.

but also the forbearance of *p* in case *q* — we have a situation that is pragmatically equivalent to the situation in which there would be neither a norm that prohibits nor a norm that commands the performance of *p* in *q*? Does the permissive norm add anything that would not be there if there merely were no norm at all, or is it pragmatically irrelevant?

2) And if it turns out that the permissive norm is not pragmatically irrelevant, is that which it adds something different from an indirect formulation or a derogation of some mandatory norms, that is, of some obligation or prohibition?

1.1. The pragmatic irrelevance of permissive norms. The category of 'permissive norms' is unnecessary. Echave-Urquijo-Guibourg (1980) and Ross (1968)

In order to clarify these questions, we will start with two significant examples of what can be called the negative answer: those given by Delia Echave, María Eugenia Urquijo and Ricardo Guibourg, on the one hand, and by Alf Ross, on the other. The exposition of Echave/Urquijo/Guibourg refers more to the first, that of Alf Ross to the second question.

"Suppose among the *charrúa* there was a group that lived from hunting and fishing and was not subject to norms or authorities of any kind. One day, they realized that other tribes attained their goals to a higher degree because of the organisation they had given themselves, and they decided to elect a leader to command them. The election fell on Toro Sentado who, in contrast to his redskin namesake [= Sitting Bull], was a peaceful Indian with little desire to interfere in the lives of other people. Thus, Toro Sentado assembled the tribe and issued his first norm: 'As of today', he said, 'it will be permitted to hunt on Tuesdays and Thursdays'. Han-Kel, an Indian with the innate talents of a shyster, tried an interpretation *a contrario*: 'Does this mean that we may not hunt on the other days?' 'Of course not', the benevolent leader hurried to clarify, 'I permit hunting on Tuesdays and Thursdays, but I'm *not* saying *anything* about the rest of the week.' Janquel was perplexed, but Onin, a tribesman with a predilection for ethical reflection, insisted: 'Does this perhaps imply a promise not to prohibit hunting on Tuesdays and Thursdays in the future?' 'No, it doesn't', Toro Sentado replied, 'I don't like to impose prohibitions on my tribe, but I do not want to exclude the possibility that I may change my mind. What ruler doesn't?' The Indians looked at each other and began to disperse in silence. They could not help feeling that the election of their leader had been, at least until this moment, totally useless. All their life they had hunted and fished as they liked, without consulting the calendar; and now, after the first law of their tribe had been issued, things would stay exactly the same, as long as it did not occur to Toro Sentado to prohibit something." (Echave/Urquijo/Guibourg 1980, 155 f.)

This little narrative very well shows the problem of the pragmatic irrelevance of permissive norms. A norm of obligation constitutes a reason for performing the action mentioned in it; a prohibitive norm constitutes a reason for forbearing that action. A permissive norm, in contrast, constitutes no reason at all, neither for performing the action nor for forbearing it. It also cannot function as a criterion for evaluating actions, since — for logical reasons — it is impossible to act in a way that is not in accordance with a permissive norm. So what difference is there between a situation where there is a norm that permits something and a situation in which there simply is no norm at all?

Still, there is a widely shared intuition that it is not the same to be permitted to do something in a certain case as not to have any norm referring to the case at all. Maybe the reason for this intuitive impression is that permissive sentences only appear to be irrelevant, that they only appear to leave the world as it was before. Actually, when a permission is given, maybe something other than what is explicitly said is done. That is the position of Alf Ross:

"Telling me what I am permitted to do provides no guide to conduct unless the permission is taken as an exception to a norm of obligation (which may be the general maxim that what is not permitted is prohibited). Norms of permission have the normative function only of indicating, within some system, what are the exceptions from the norms of obligation of the system." (Ross 1968, 120)
"I have never heard of any law's being passed with the purpose of declaring a new form of behaviour (e. g., listening to the wireless) permitted. If a legislator sees no reason to interfere by issuing an obligating prescription (a command or a prohibition) he simply keeps silent. I know of no permissive legal rule which is not logically an exemption modifying some prohibition, and interpretable as the negation of an obligation." (Ibid., 122)

And about von Wright's idea — to which we shall return presently — of understanding the "constitutional guarantee of certain freedoms" of the citizen as promises of non-interference by the legislator, he says:

"The idea of a promise made by a legislator to the citizen, creating a moral obligation which binds the legislator, is a figment of the imagination and has long since been abandoned in legal theory. The constitutional guarantee of certain freedoms has nothing to do with promises, but is a restriction of the power of the legislator, a disability which corresponds to an immunity on the part of the citizen. The legislator does not promise not to use a power which he possesses, but, rather, his power (or competence) is defined in such a way that he *cannot* legally interfere with the liberties guaranteed. Any legislative act to this effect would be unconstitutional and therefore null and void." (Ross 1968, 123 f.)

At least in two respects, this conception deserves a more detailed examination. The first is that even if it would be acceptable to understand the constitutional guarantee of certain freedoms as defining spheres of legislative 'disability' or incompetence, this could apply only to rigid constitutions, but not to the provisions of flexible constitutions. The second is that — for reasons we have already explained in ch. II and to which we return here — to understand constitutional guarantees as spheres of legislative incompetence is clearly unconvincing. We will look at both questions in turn.

Let's begin with the case of flexible constitutions, meaning that according to their own terms "they may be modified by the ordinary legislative organ through the ordinary procedures of law-giving" (Guastini 1993, 72). In that case, the constitution-giving norm authority is not placed above the ordinary legislator: it is the ordinary legislator himself. Thus, the constituent body can neither confer nor restrict the normative power of legislating, because the relations between the constitution and the law are regulated simply on the basis of the principle of *lex posterior*. A later law that introduces, for example, norms of prohibition where the constitution stipulated permissions would be a perfectly regular normative change rather than an unconstitutional law.

So the applicability of the idea that constitutional guarantees of certain freedoms are equivalent to spheres of legislative incompetence seems to be limited to those cases where the constituent authority is different from, and placed above, the ordinary legislative authority, i. e., to rigid constitutions. But even here, the idea is unconvincing, for the following reason: that a norm is legally 'null' can mean two radically different things. It can mean, in the first place, that that norm is not recognized as such by the legal order, i. e., that from the perspective of that order it simply does not exist as a norm; and it can mean, secondly, that the legal order recognizes it as a norm, but imposes on some organ the duty to nullify it. We can think, for instance, of a 'law' with a perfectly constitutional content, but issued by a private individual. That 'law' is 'unconstitutional' or 'null' in the sense that it is not recognized by the legal order as a 'law'; from the perspective of the legal order, that 'law' simply does not exist. Or think of a statute approved by the constitutionally required parliamentary majority, but whose content conflicts with constitutional norms, e. g., because it does not respect permissions given to the citizens by the constitution. Here, the situation is totally different: that statute is recognized by the legal order as a statute, although. if an authorized organ challenges its constitutionality before the constitutional court, that court has the duty to nullify the statute

— where that annulment obviously implies a reproach for the legislator. We can say that in the first case the private individual, while having respected the constitutional norms imposing certain prohibitions concerning the possible content of statutes, simply has not (and could not have) successfully used the constitutional norm conferring the power to legislate on the parliament; whereas in the second case, parliament has used that norm successfully and, therefore, produced the intended normative change (has issued the statute in question), although the Constitutional Court has the duty to nullify its result because it violates prohibitions concerning the exercise of that normative power. That means that the constitutional guarantees of certain rights and liberties do not imply spheres of incompetence of the ordinary legislator (because if that were the case, the statute of unconstitutional content issued by parliament simply would not exist as a statute, just as the statute issued by the private citizen, whether its content is constitutional or not, does not exist), but only prohibitions to exercise that competence in order to produce statutes with certain contents.

Although it is a correction of Ross, this conclusion supports his central thesis, namely, that there are no purely permissive norms, that permission is not an independent modality, but only the exception to (or the derogation of) mandatory norms (of obligation or of prohibition) or the indirect formulation (addressed to subjects other than their explicit addressees, i. e., to lower norm authorities) of mandatory norms (of obligation or prohibition).

1.2. Von Wright's proposal: Permissive norms as promises

Von Wright's proposal to understand permissions as promises by those who issue them deserves more attention than Ross' remarks quoted above may suggest. The conception of permissions as promises is the final result of the careful analysis of permissions (or, more precisely, of 'permissive prescriptions') carried out in *Norm and Action* (von Wright 1963), on which we will concentrate our attention.[2]

In *Norm and Action*, von Wright examines the two ways or forms in which "it has been attempted to deny the independent status of permissions: The one is to regard permissions as nothing but the absence or non-existence of 'corresponding' prohibitions. The other is to regard permissions as a peculiar kind of prohibition, *viz.* prohibitions to interfere with an agent's freedom in a

---

[2] On the meanderings of his reasoning and on the different positions von Wright has taken on permissions in the course of his work, cf. Daniel González Lagier (1995).

certain respect" (von Wright 1963, 85). His analysis of the first of these two ways leads von Wright to formulate the distinction between weak and strong permission, while examination of the second makes him characterize 'permissive prescriptions' as promises. We will look at both of them in turn.

Von Wright thinks that the simple identification of the permission to do a certain thing with the absence of a prohibition to do that same thing is 'in serious error', for the following reason:

"One cannot make an inventory of all conceivable (generic) acts. New kinds of act come into existence as the skills of man develop and the institutions and ways of life change. A man *could not* get drunk before it had been discovered how to distil alcohol. In a promiscuous society there *is* no such thing as committing adultery.
As new kinds of acts originate, the authorities of norms may feel a need for considering whether to order or to permit or to prohibit them to subjects. [...]
It is therefore reasonable, given an authority of norms, to divide human acts into two main groups, *viz.* acts which are and acts which are not (not yet) subject to norm by this authority. Of those acts which are subject to norm, some are permitted, some prohibited, some commanded. Those acts which are not subject to norm are *ipso facto* not forbidden. If an agent does such an act the law-giver cannot accuse him of trespassing against the law. *In that sense* such an act can be said to be 'permitted'.
If we accept this division of acts into two main groups — relative to a given authority of norms — and if we decide to call acts permitted simply by virtue of the fact that they are not forbidden, then it becomes sensible to distinguish between two kinds of permission. These I shall call *strong* and *weak* permission respectively. An act will be said to be permitted in the weak sense if it is not forbidden; and it will be said to be permitted in the strong sense if it is not forbidden but subject to norm. Acts which are strongly permitted are thus also weakly permitted, but not necessarily vice versa.
Roughly speaking, an act is permitted in the strong sense if the authority has considered its normative status and decided to permit it." (von Wright 1963, 86)

Now, if one accepts von Wright's distinction between weakly and strongly permitted acts,[3] the following three questions arise: *1)* Does the introduction of a

---

[3] With respect to that same passage in von Wright, Alchourrón and Bulygin have pointed out that it contains two different definitions of *strong permission*: *i)* an act is strongly permitted "if it is not forbidden but subject to norm"; *ii)* an act is strongly permitted "if the authority has considered its normative status and decided to permit it". They comment this as follows: *(ii)* is apparently meant to be a complementary explanation of *(i)*, but in fact *(i)* and *(ii)* are not identical and so the question arises as to which of the two is to be regarded as a definition of 'strong permission'. Indeed, *(i)* requires two conditions: *(a)* an act must not be forbidden, and *(b)* it must be 'subject to norm', i. e. it must be either permitted or obligatory. *(ii)*, instead, requires only one condition: it must be (explicitly or implicitly) permitted. This second condition is identical to *(b)*" (Alchourrón/Bulygin 1984, 351). Obviously, definition *i)* eliminates the possibility of strong permissions conflicting with prohibitive norms, and this elimination is not justified. Besides, an

norm that permits act *X* in any way change the status of act *X*, previously not subject to a norm? As we have just seen, von Wright says that if an agent does (or, we may add, forbears) a weakly permitted act, "the law-giver cannot accuse him of trespassing against the law". But the situation is exactly the same if he does (or forbears) an action doing and forbearing of which are strongly permitted, for the simple reason that permissions can neither be complied with nor not complied with. *2)* Does the introduction of a permissive norm with respect to act *X* change the status of other acts different from, but related to *X*, that is, acts of hindering or sanctioning the performance of *X*? *3)* Does this change in the status of other acts different from the one the permissive norm explicitly mentions imply something more than the prohibition to hinder or sanction those acts? If the answer to the first question is negative, then the possible relevance of permissive norms would have to be found in relation with other acts different from those the norms explicitly mention. And if the question to the third question also is negative, then so-called permissive norms are nothing but an indirect formulation of prohibitive norms.

But in order to see how von Wright answers those questions, we first need to see how he treats the second of the above-mentioned ways of denying an independent status to permissions, that is, that of regarding them as prohibitions to interfere. In that respect, von Wright notes that "It seems possible to distinguish between various kinds of strong permission" which are — in an order of increasing strength — permission as *toleration*, permission as a *right*, and permission as a *claim*:

"In permitting an act the authority may only be declaring that he is going to *tolerate* it. The authority 'does not care' whether the subject does the act or not. The authority is determined not to interfere with the subject's behaviour as far as this act is concerned, but he does not undertake to protect the subject from possible interferences with his behaviour on the part of other agents.
Any (strong) permission is at least a toleration, but it may be more than this. If a permission to do something is combined with a *prohibition to hinder or prevent* the holder of the permission from doing the permitted thing, then we shall say that the subject of the permissive norm has a *right* relatively to the subjects of the prohibition. In granting a right to some subjects, the authority declares his toleration of a certain act (or forbearance) and his intolerance of certain other acts. [...]
We ought to distinguish between *not* making an act *im*possible (for someone to perform) and *making* an act *possible*. The second is also called *enabling* (someone to do something). It is the

---

important difference between strong and weak permissions seems to be that the former, because they depend on a permissive norm, can give rise to contradictions in the system (between the permissive norm and an eventual prohibitive norm), which, of course, cannot happen in the case of weak permissions (since they are defined merely as the absence of a norm).

stronger notion. Enabling entails not-hindering, but not-hindering does not necessarily amount to enabling.
If a permission to do something is combined with a *command to enable* the holder of the permission to do the permitted thing, then we shall say that the subject of the permissive norm has a *claim* relatively to the subjects of the command. It is understood that any claim in this sense is also a right, but not conversely." (von Wright 1963, 88 f.)

A 'permission as a right' is a 'permission as toleration' plus a prohibition to hinder or prevent performance of the permitted act; a 'permission as a claim' is a 'permission as a right' plus a command to make performance of the permitted act possible. What 'permissions as rights' and 'permissions as claims' add to 'permissions as tolerations' are, respectively, prohibitions and obligations concerning the behaviour of others. As von Wright himself says, "the specific characteristics of the [last] two species of (strong) permission [...] can be accounted for in terms of prohibition and/or command". And thus, he continues, "if there is an element in permissions which is not reducible to the other norm-characters this element is identical with what we called *toleration*" (ibid., 90). "In order to see whether permission is an irreducible norm-character or not, we must thus examine the notion of toleration." A declaration of toleration can be either a declaration of intention — which does not express any norm, because "a declaration of intention is not a normative concept at all" — or a promise — and then "the question whether permission is definable in terms of the other norm-characters would be reducible to the question whether the normative character of a promise (or at least of a promise of non-interference) can be accounted for in terms of 'ought' and 'must not'" (ibid., 91). In von Wright's view, "That the answer to the last question is affirmative would probably be universally conceded" (ibid.). Therefore, although our author ends his treatment of permissions saying that "On the question whether permission is or is not an independent norm-character, I shall not here take a definite stand" (ibid., 92), it looks as if from his analysis we can draw the negative conclusion, i. e., that what in permissions is not reducible to prohibitions or commands to others can be reduced to self-prohibitions by the issuing authority.

In any case, in von Wright's opinion, that last notion is problematic, because "on the view of permissions as promises, permissions would be self-reflexive prescriptions, *viz.* self-prohibitions". And prescriptions "require an authority and a subject" which, in the case of permissions as toleration, would be one and the same. Whether this is possible, whether self-reflexive prescriptions are possible, is, in von Wright's view, open to question, and therefore, "If we think that such prescriptions cannot exist we should have to conclude that permissions are not prescriptions" (ibid., 91). But that would lead one to regard permissions as norms of a moral kind, because "That promises ought to be kept would ordinarily be thought of as a typically *moral* norm, and the obligation to do this or that because one has promised to do it would be called a moral obligation" (ibid., 92).

This last assertion is, of course, questionable, not only because, if 'Promises ought to be kept' is a moral norm, then, in the last instance, 'Prescriptions of an authority ought to be complied with' also would be one, but for reasons more internal to the law. Because the notion of 'self-prescription' may be problematic; but, of course, the notion of a 'voluntarily created obligation', as the consequence of a promise, does not seem problematic at all, since this type of obligations are the basis of such uncontrovertibly legal institutions as contracts as well as unilateral acts-in-the-law in which — as there is no exchange of promises but only a promise in one direction — the situation is substantially the same as in promises as toleration, as presented by von Wright. That an authority creates an obligation for itself does, in our opinion, not seem to produce any special problem for legal theory.

We do think, however, that there are a few observations worth to be made before concluding the analysis of von Wright's treatment of permissions. The first is that, regardless of von Wright's final reservations, conceiving of permissions as promises of non-interference leads one to deny that permissions are an independent category of norms: Where they cannot be reduced to prohibitions to others, they are nothing but self-prohibitions by the respective authority. The second observation is that conceiving of permissions as promises of non-interference (and, therefore, as self-prohibitions) seems to be especially suited for permissions issued in more 'informal', less institutionalized contexts than the legal one. If, for example, a father says to his son that he has permission to go play outside, we can readily agree that the father thus acquires the obligation not to punish his son if he does go outside to play. In legal contexts, however, that sense seems to be much more limited. The constituent assembly never prohibits *only* itself to interfere, for instance, in the freedom of expression. Actually, there would be a self-prohibition only if we were dealing with a constitution that prohibits its own modification in that point (which is rather infrequent and creates problems of a very different kind than those we are treating here); but even in that case, there would be a prohibition of interference for someone else, namely, first of all, for the ordinary legislator who is usually the most important addressee of that kind of constitutional prohibitions. The same occurs when the ordinary legislator permits something: prohibitions of interference with the permitted conduct are not addressed to himself (the principle of *lex posterior* seems to be practically universal in the legislative sphere), but to executive and judicial organs as well as to the general public. Thus, Ross seems to be right when he says about permission as toleration that "I know of no legislative act which says this" (1968, 123).

So, whether we regard permissions as the absence of prohibitions or as prohibitions of interference, the analysis of von Wright's work brings us to the conclusion that both ways lead one to deny independent status to permissions.

1.3. Weak and strong permission in Alchourrón and Bulygin

Of all important works in contemporary legal theory, the one which most stresses the relevance of the distinction between strong and weak permission is probably that of Alchourrón and Bulygin.[4] As the reader may know, that distinction is one of the cornerstones of *Normative Systems* (1971) — the book with which, in von Wright's words, Alchourrón and Bulygin "stepped on the international stage" —, and it also plays a central role in later papers like 'The expressive conception of norms' (1981), 'Permission and permissive norms' (1984) or 'Libertad y autoridad normativa' (1985) (all of them reprinted, in Spanish, in Alchourrón/Bulygin 1991). Of these, 'Permission and permissive norms' seems to be the most complete elaboration of their position on the subject. We will now try to give a faithful description of their basic theses. From the outset, we want to point out that in our view, and contrary to what the authors themselves insist on, from these theses one can infer that the categories of 'prescriptive permission', 'strong permission' and 'permissive norm' are superfluous. Now, to the theses themselves.

1) The first refers to concepts of permission. According to Alchourrón and Bulygin, we must distinguish three of them: a prescriptive concept of permission, plus the descriptive concepts of strong and weak permission:

"When the term 'permitted' occurs in a norm it expresses the *prescriptive concept of permission* [...] But the same term occurring in a norm proposition is ambiguous: when one says that a state of affairs p is permitted by a set of norms α, this may mean two different things: either that there

---

[4] Rafael Hernández Marín (1993) has also written about that distinction; he has presented some criticism with respect to the symbolization used by the Argentine authors, their definitions of strong and weak permission, and the relationship between these two kinds of permissions, which, however, do not affect our general line of reasoning.

is a norm (in α) permitting that *p*, or that *p* is not prohibited by α. So there are *two concepts of descriptive permission: strong permission* [...] and *weak permission*" (Alchourrón/Bulygin 1984, 353).

2) In contrast to what happens in complete and consistent normative systems, where the distinction between weak and strong permission "vanishes [because] both concepts overlap" (ibid., 353), in the context of normative systems that are incomplete (where there are conducts that are permitted in the weak sense, but not in the strong sense) or inconsistent (where there are conducts permitted in the strong sense, but not permitted in the weak sense, because the system also contains a norm that prohibits them) that distinction acquires special relevance (ibid.).

Up to this point, we have no objection against Alchourrón/Bulygin's theses. That some behaviour may be, at the same time, permitted in the strong sense and prohibited by some other norm of the same system means no more than that normative systems can contain antinomies; and that is, in our view, something that cannot be denied. But this alone is not a sufficient argument for regarding what they call prescriptive permission to be an independent norm-character, because that same situation could also be described — without having to refer to permissions — as an antinomy between two norms, one of which prohibits a certain conduct in a certain case while the other negates that prohibition (i. e., has the form '*p* is not prohibited in case *q*').[5] In order to be able to regard prescriptive permission as an independent norm-character, it would have to be shown *a)* that when a behaviour is covered by a prescriptive permission (that is, by a permissive norm which allows us to say, at the level of norm-propositions, that the behaviour in question is strongly permitted), the normative status of that behaviour is different from when it is only weakly permitted; and *b)* that the modification of the system produced by the introduction of a permissive norm is different from the negation or nullification of prohibitions as well as from the introduction, by an indirect formulation, of prohibitions of behaviour different from, though related to the behaviour mentioned in the permissive norm (that is, from behaviour consisting in prohibiting, hindering or sanctioning the behaviour mentioned in the permissive norm). As they them-

---

[5] You might say that "*p* is not prohibited in case *q*" is a strange formulation of a norm, and that is certainly true. But: *1)* here, 'strange' only means 'shocking', in the sense of not being part of common stylistic usage, rather than 'nonsensical' or 'containing some logical defect'; *2)* also, Alchourrón and Bulygin offer examples for very similar norm formulations, as, e. g., when they speak of "a general *norm* of the form '-*Op/q*'", or of "a norm [...] expressed by '*P-p/q*' or '-*Op/q*' or '-*Ph-p/q*'" (Alchourrón/Bulygin 1971, 161 f.).

selves write, the crucial problem is contained in the following question: "which is, after all, the practical difference between strong and weak permission, i. e. between permitted and simply not forbidden action" (ibid., 368).

3) Now, according to what Alchourrón and Bulygin themselves say (although probably not according to what they mean to say), that 'practical difference' between weak and strong permission, that is, the modification produced in a system by introducing a permissive norm, can be explained completely in terms either of the negation or cancellation of prohibitions, or by an indirect formulation of prohibitions.

3.1) Slightly reformulating an example of Alchourrón and Bulygin themselves (1971, 165), suppose we have the following two norms:

N_1: In circumstances A and B, p is prohibited.
N_2: In circumstances non-A and non-B, p is permitted.

Now, continuing with the example, imagine a state of affairs such that circumstances A and non-B hold. According to Alchourrón and Bulygin, in such a situation, the interpreter cannot find a satisfactory solution, because "[t]he *argumentum e contrario* allows us to infer two incompatible conclusions, according to which of the two norms is adopted as a premiss". Thus, the solution of the problem must come from the introduction of a third norm:

N_3: In circumstances A and non-B, p is permitted.

This norm allows us to solve the case without having to decide beforehand whether N_1 or N_2 should be adopted as the premise of an *argumentum e contrario*. Now, obviously, that solution could also be reached (without any need to introduce a permissive norm) if instead of N_1 we had:

$N_{1'}$: Only if circumstances A and B are given jointly, p is prohibited.

That means: N_3 only limits (by specifying it for an unforeseen case) the scope of the prohibitive norm N_1. N_3 negates the extension of the prohibition stipulated in N_1 to the case consisting in circumstances A and non-B. If N_1 is replaced by $N_{1'}$, N_3 as well as N_2 are superfluous.

3.2) An important function of permissive sentences is that of derogating prohibitive norms.

"[A] prohibition certainly cannot be lifted by means of another prohibition. In order to cancel or derogate a norm of obligation we need to perform another kind of normative act, which is radically different from the act of issuing a prohibition. Permissive norms often (if not always) perform the important normative function of derogating prohibitions." (Alchourrón/Bulygin 1984, 368)

But the kind of normative act necessary for derogating a prohibitive norm does not necessarily have to be understood as an act of permission. It can be regarded simply as an act of derogation (that is, of canceling a prohibition). If by understanding it in this way we can explicate the same thing then, by the principle of economy, this alternative is preferable since we do not need to introduce permission as an independent norm-character. That is precisely what Alchourrón and Bulygin admit in 'The expressive conception of norms'. Under the heading 'Permission', they ask how an expressivist theory of norms can account for acts consisting in giving permission to perform an action p, and they say that

"There seem to be two possible ways out of this difficulty. (i) One way is to describe this act as an act of lifting a prohibition, i. e. as derogation of the prohibition of p. (ii) An alternative way is to accept a new kind of normative act, the act of giving or granting a permission (for short: act of permitting). If this is accepted, then it must also be accepted that there are two kinds of norms, mandatory norms and permissive norms (in the sense in which an expressivist uses the term 'norm') [...] At this state, one feels tempted to ask: are there really two distinct analyses? What is the difference, if any, between promulgating a permission and derogating a prohibition? [...] One has the impression that both analyses are substantially equivalent in the sense that they are two different descriptions of the same situation. If this were so, it would be a rather surprising result; it would show the fruitfulness of the concept of derogation and its importance for the theory of norms. The concept of a permissive norm could be dispensed with; a fact that would justify the position of those expressivists that only accept mandatory norms, provided they accept the existence of derogation." (Alchourrón/Bulygin 1981, 116 f., 118, 119)

3.3) Still another important function of permissive norms is regulating the exercise of the normative powers of authorities of lower rank than the one issuing the permissive norm. In 'Permission and permissive norms', they write with respect to the narrative of Echave/Urquijo/Guibourg quoted earlier:

"Let us suppose that one day Toro Sentado decides to appoint a minister. The minister is authorized by him to issue new norms regulating the behaviour of the people and to derogate them, but he has no competence to derogate the norms issued by Toro Sentado himself. In this case the permission given by Toro Sentado to hunt on Tuesdays and Fridays functions as a limitation of the competence of his minister: the minister cannot derogate these norms and so he cannot prohibit to hunt on those days, though he can prohibit to hunt on any other day of the week. So such permissions can be interpreted as a rejection in advance of the corresponding prohibitions" (Alchourrón/Bulygin 1984, 369 f; cf. also Alchourrón/ Bulygin 1985, 244 ff.)

As the reader will have noticed, in this text Alchourrón and Bulygin think that prohibitions of the exercise of some competence are part of the competence norm itself. In our view — for the reasons pointed out with respect to Ross — it is preferable to see such prohibitions as the content of regulative norms about the exercise of competence, i. e., regulative norms that are different from the competence norm itself. However, for two reasons, this is not very important for our present context: First of all, because the distinction between the norm of competence and the regulative norms guiding its exercise has later been accepted by the authors we are examining here;[6] thus, in 'On Norms of Competence', Eugenio Bulgin writes that "[s]ituations in which a person has legal competence to produce certain types of acts and at the same time is forbidden to make use of his competence are relatively frequent" (Bulygin 1992, 205 f.). Secondly, because the question of whether or not prohibitions to exercise a competence are part of a competence norm itself is not essential for the topic we are interested in; what is important is that, according to Alchourrón and Bulygin, permissions given by a higher-ranking authority to a lower-ranking authority are nothing but indirectly expressed prohibitions. Therefore, from this perspective too, there is no reason for regarding permission as an independent norm-character.

2. Reformulating the problem

2.1. Permission and the regulation of 'natural' conduct

So far, we have examined what we believe to be the most relevant statements to be found in contemporary legal theory on the topic we are interested in. Now, we must come back to the two questions formulated at the beginning and try to confront them directly. The first was whether the existence of permissive sentences (making it facultative to do or omit some action p) is irrelevant, that is, whether or not it implies any changes, as compared to a situation where the

---

[6] There are two phases in Alchourrón/Bulygin's treatment of norms of competence. In the first, they regard them as permissive norms; later, they see them as definitions or conceptual rules. For a critique of both perspectives, cf. ch. II.

legal system in question does not contain any norm referring to p. The second question — arising only when the answer to the first question is not totally negative — was what the relevance of such permissive sentences could be and whether or not it can be expressed in terms of mandatory norms (norms of obligation or prohibition), that is, without the need for introducing the category of permissive norms.

That the answer to the first question is not completely negative can be seen more clearly when one realizes that the question is somewhat ambiguous. Because to say that in a legal system S there is no norm referring to p can mean that

1) p is not a conduct falling within the sphere regulated by S; in other words, conduct p is indifferent with respect to S;

2) p is a conduct that is relevant for system S, but there is no norm referring to p, because:

a) such a norm has not been considered necessary (for example, because the possibility that extending the scope of some mandatory norm to include p would be a sustainable interpretation or that a lower-ranking authority may try to introduce a mandatory norm referring to p has not been considered), or

b) conduct p has not been foreseen:

b') for subjective reasons, that is, through a fault of the legislator;

b") for objective reasons, that is, because after the norms of system S were enacted, new circumstances — new possibilities of action — have arisen which the legislator could not foresee.

Obviously, the distinction between *1)* and *2)* presupposes that there is such a thing as legally indifferent behaviour, that is, behaviour a legal system is just not interested in, at least not for the time being. Thus, the concept of legally indifferent behaviour is relative to a specific legal system at a specific point in time. Of course, it may be difficult to determine where the legally indifferent ends, at a certain time and with respect to a certain legal system, i. e., to determine (in the terminology of Alchourrón/Bulygin 1971) the Universe of Discourse (the set of situations and states of affairs) and the Universe of Actions a specific legal system is interested in. But, in any case, the concept of indifferent behaviour is, in our opinion, clearly applicable to partial subsystems (which lawyers usually work with) such as, for example, family law, or the law of the autonomous community of Valencia. In those cases, the Universe of Discourse and the Universe of Actions may have some zones of penumbra, but that implies that there is also a zone of clarity: it seems obvious that there are forms of behaviour family law, or the law of the autonomous community of Valencia, does not care about. With respect to the legal system as a whole, however, it is doubtful whether there is any truly indifferent behaviour. The fact that, as Raz (1990, 154) says, legal systems claim authority for regulating any kind of behaviour, i. e., they claim materially unlimited competence,[7] is a good argument in favor of the thesis that (from the point of view of a legal system as a whole) there is no indifferent behaviour. Here, one could say that — because of the materially unlimited competence legal systems as such claim — whatever is not prohibited is permitted, that is, legally regulated.

Now, regardless of whether or not one considers it a legally indifferent action, it would not make much sense if, for instance, the board of the University of Alicante would issue a norm permitting professors to choose the colour of their trousers. In that case, the addressees would probably think that it is a joke or that something worrying had happened to the mental health of the members of the board. From the point of view of the practical deliberation of the addressees, such a norm would leave things exactly as they were, for the simple reason that before, everybody had thought that choosing the colour of one's trousers was everyone's own business, just as it is after the norm is issued.

If one assumes, as we think one should, that a norm must in some sense guide the conduct of its addressees, that would mean that a permissive norm can only perform that function — i. e., will not be superfluous — if it is issued in a context in which there is one of the circumstances earlier mentioned in 2), which we have called C_2, or else a new circumstance C_3, consisting in the fact that the corresponding legal system S had until now regulated conduct p through a mandatory norm, that is, had stipulated that the behaviour in question was either obligatory or prohibited.

The issuing of a permissive norm in system S — say, at time t_2 — changes things as compared to that same system at time t_1, in one of the following two senses:

1) If at time t_1 circumstances C_2 held, then the change consists in clarifying, or determining, the normative status of p; since there was no norm explicitly saying that p was a facultative behaviour, by clarifying the situation — and by prohibiting that lower-ranking authorities introduce prohibitive norms with respect to p — the new norm gives security to the addressees and, in that

---

[7] For a critical assessment of this thesis, however, see ch. V, sect. 7.2.

sense, perfectly well contributes to guiding their behaviour: the fact that they now know with certainty that a certain behaviour is facultative — that is, free of normative restrictions — undoubtedly will lead many of them to do (or to stop doing) what they would otherwise not do (or not stop doing).

2) If at time t_1 circumstance C_3 held, then the issuing of a permissive norm at t_2 changes things in the sense that the deontic status of p is modified: from being obligatory or prohibited, it changes to being facultative. In normative terms, the relevance of the new norm is obvious: it frees behaviour p from the normative restrictions it was subject to before.

The answer to the second question we posed at the beginning is that the 'relevance' of permissive norms — or sentences — apparently can be explained completely in terms of the negation or derogation of, or the exception from, mandatory norms (or in terms of indirect formulations of such norms), and, possibly, of definitions.

Imagine two situations corresponding to C_2. The first is the following: at t_1, in the system there is a norm prohibiting indecent behaviour at the beach; the question is whether this includes women going topless. The issuing, in t_2, of a norm saying that 'It is permitted for women to be topless on the beach' is pragmatically equivalent to the issuing of a (partial) definition of 'indecent behaviour' in the sense that 'Being topless is not considered indecent behaviour'. The second situation corresponding to C_2 would be as follows: in t_1, there is no norm referring to the clothes to be worn on the beach, but there is rumour that some local authorities intend to hinder, prohibit and/or sanction women being topless. If Parliament then issues a norm saying that 'It is permitted for women to go topless on the beach', that would be pragmatically equivalent to the issuing of a prohibitive norm addressed to the local authorities (and, in general, to all authorities of a lower rank than Parliament) in the sense that 'It is prohibited to hinder, prohibit and/or sanction going topless'.

Now, let us look at a situation of the kind of C_3. At t_1, the system contains a norm prohibiting to take horses out of the province they come from. Obviously, the issuing of a permissive norm — 'It is permitted to take a horse out of the province it comes from' — is pragmatically equivalent to the issuing of a derogating provision — 'The norm that prohibited to take a horse out of the province it comes from is derogated'. Similarly, the issuing at t_2 of a norm in the sense that 'It is permitted to take a horse from one province of Andalucia to another' would be pragmatically equivalent to the issuing of a provision excluding the provinces of Andalucia from the prohibition established at t_1.

2.2. Permission and the exercise of normative powers

As one can easily see, the examples of permission we have been using belong to the sphere of rules regulating *natural behaviour* (where this means behaviour that does not consist in the exercise of a normative power). We will now go on to look at the regulation of the exercise of normative powers. By this, we mean powers conferred by rules whose canonical form, as we already explained in the previous chapter, is of the type 'If state of affairs X obtains and Z performs action Y, then institutional result (or normative change) R is produced'. A rule of this kind can, in turn, be affected by other rules, with respect to action Y or to result R. We will start with the latter.

In a very general characterization, we can distinguish between normative powers whose exercise is obligatory (as, for example, judicial power: judges have the obligation to produce institutional results of the 'judgment' type) and normative powers whose exercise is facultative (as, for example, the powers referring to the conclusion of private contracts).[8] But this distinction is too rough, because a result could also be partly obligatory and partly facultative. For instance, according to the Spanish legal order, the government has the obligation to propose to the king a candidate for the appointment of prosecutor general, but it is relatively free in its choice of a candidate (relatively, because the choice must be from "among Spanish lawyers of recognized renown,

---

[8] There may even be normative powers whose very exercise is prohibited to the power-holder himself. In one of the papers already quoted, E. Bulygin writes that "[s]ituations in which a person has legal competence to produce certain types of acts and at the same time is forbidden to make use of his competence are relatively frequent and are not regarded as conflictive. Take the following example. According to Argentine law, a petition directed to a court must be signed by a professional lawyer *(abogado)* in order to be legally valid. Only lawyers have competence for it. Now if a lawyer is appointed a judge (and in order to be a judge one must be a lawyer) then he is no longer allowed to act as a professional lawyer, for judges cannot represent parties before a court. Suppose now the following situation: a judge A is asked by a friend to sign a petition. A is a lawyer, so he has competence to sign the petition. This means that the petition signed by A would be a valid petition and treated as such be the courts. But as he is a judge he is forbidden to sign the petition. This means that in doing it he would be violating one of his duties as a judge, resulting in a liability" (Bulygin 1992, 205 f.). A similar example in the field of canon law is that of the exercise of the normative powers reserved to catholic priests and bishops (sacramental powers) by priests and bishops subject of a suspension *a divinis*. However, that the exercise of a normative power is prohibited to the holder of that same power is rather rare, because — apart from a magical (for example, a sacramental) conception of normative powers — it does not seem to make much sense to uphold someone as the holder of a normative power while, at the same time, totally prohibiting its exercise to that same person. Another question — to which we have already referred earlier in the text — is that some of the normative results the holder of a normative power can (i. e., has the normative capacity to) produce may be a normative result that is prohibited by regulative norms guiding the exercise of that power (for example, when an illegal sentence or an unconstitutional law is issued). On these problems, cf. above, ch. II.

having effectively exercised their profession for more than fifteen years"). So, for the government it is obligatory to propose a Spanish lawyer of recognized renown, etc., to be appointed as prosecutor general by the king; but it is facultative for it to propose, for example, the Spanish lawyer, of recognized renown etc., Mr. Garcigómez. It may also be the case that the exercise of one and the same normative power is obligatory in some cases and facultative in others. According to Spanish law, for example, the members of the administrative board of a company are the holders of the normative power to convene a general meeting, and such a convention is obligatory in certain cases (in certain time intervals, when shareholders representing a certain percentage of total capital so request, etc.), and facultative in all others.

In order to clarify what it means that a normative result is 'facultative', remember the differences between mandatory and power-conferring rules as reasons for action. For this purpose, we will make use of Joseph Raz's theory of rules as reasons for action (1990), on the one hand, and a translation of some kinds of Kantian imperatives into the language of reasons for action, on the other. According to Raz's theory, mandatory rules are operative reasons of a special kind: they are protected or peremptory reasons for doing what they stipulate, excluding one's own judgment about the reasons pro and con as a guide for conduct. If, instead of Raz's theory, we adopt the Kantian classification of imperatives and adapt it to the language of reasons for action, we can say that they are categorical reasons: they indicate what should be done, regardless of the subject's wishes and interests. A permissive sentence can thus be regarded as the negation of a peremptory or categorical reason. Such sentences help to guide action, we said, insofar as, by cancelling or clarifying the non-existence of such a peremptory reason (for doing or omitting *p*), they make it possible for the wishes and interests of the subject in question to function as operative reasons.

Power-conferring rules, in contrast, affect behaviour in a very different, indirect way. They indicate how some given end, consisting in a certain normative result, can be attained. Thus, they are not — in Raz's terminology — operative reasons, but auxiliary reasons. Or, translating the Kantian classification of imperatives into the language of reasons for action, we could say that they are not categorical, but hypothetical reasons. What happens here is that the legal order assumes that some of those ends or results the subject may want or not want — for example, to become a married person, to sell a rural estate belonging to him, to convene a shareholders' meeting if doing so is not obligatory

—, while others are results the subject simply cannot not want: a judge cannot not want to issue a decision; the government cannot not want to appoint a prosecutor general, etc. One could draw a parallel to what Kant called problematic hypothetical imperatives and assertoric hypothetical imperatives and, consequently, speak of problematic hypothetical reasons and assertoric hypothetical reasons. A power-conferring rule is always a hypothetical reason, because it provides a reason for doing something if a certain state of affairs occurs, provided one also has a reason for attaining a certain result or end. A hypothetical reason is problematic if the power-holder is free to decide whether or not she has that reason for attaining the result; and it is assertoric if the reason for attaining the result is imposed on the power-holder, that is, if it escapes his (normative) control. That the normative result is regulated as facultative, then, becomes the 'indicator' showing that a power-conferring rule operates as a problematic hypothetical reason. Its function, thus, is that of specifying what kind of hypothetical reason the power-conferring rule it belongs to actually is. But, as we have already pointed out, a normative result may be obligatory in some of its elements, and facultative in others, or it may be obligatory under certain circumstances and facultative under others. If an element of the result (for example, to propose a candidate for the appointment of prosecutor general, and to do this by naming a Spanish lawyer of recognized renown, etc.) is obligatory while another one (e. g., to propose a specific person) is facultative, the rule conferring the power to produce that result is an assertoric hypothetical reason with respect to the first, and a problematic hypothetical reason with respect to the second element. If bringing about the result is obligatory in some cases and facultative in others, the power-conferring rule is an assertoric hypothetical reason in the former, and a problematic hypothetical reason in the latter cases.

Concerning rules regulating natural conduct, we have seen before that the circumstances in which it makes sense to issue a permissive norm are either that there is some doubt about whether or not a given mandatory rule is applicable to the behaviour in question (what we called C_2), or that there exists a mandatory norm whose applicability to that conduct one wishes to cancel (what we called C_3). Are these the same circumstances that give sense to the issuing of permissive norms in the context we are discussing here? Obviously, when the permissive norm is issued simultaneously with the rule conferring the corresponding power, it makes no sense to speak in terms of C_3, for the simple reason that the form of behaviour consisting in bringing about a normative result has as a condition for its possibility the rule conferring the corresponding normative power. But it is, of course, perfectly possible that bringing about a certain normative result is modalized as obligatory (or prohibited) at the time the corresponding power-conferring rule is issued, and that one later wishes to cancel or restrict the scope of that mandatory rule. In order to do this, one can — as in the case of natural conduct — either issue a derogating provision or a permissive rule. With respect to circumstances C_2, from what we said before, it is obvious that, when the permissive rule about the exercise (or an element of the exercise) of a normative power is issued simultaneously with the rule conferring the power in question, the issuing of the former cannot fulfil the function of clarifying a *previously existing* doubt, but rather that of preventing *ab initio* that such a doubt arises, by formulating, at the very moment the normative power is conferred, a negation of the applicability of some mandatory rule, i. e., some peremptory reason, to its exercise (always or under certain circumstances, concerning that power as a whole, or some of its elements).

Let us now examine the action the power-conferring rule links to the production of the normative result in question, that is, to element Y of our 'canonical form'. As we already indicated in the appendix to the last chapter, speaking of an 'action' in this context is a simplification. Because, as we said then, power-conferring rules usually do not link the production of a normative result to one single action but rather either to some set of actions (a course of action), to a disjunction of courses of action, or to a combination of both. In this context, we said, to qualify a course of action (a certain set of actions) or some fragment of it as obligatory means that one must necessarily follow that course of action, or fragment of it, in order to bring about the normative result (that is, that course of action or that fragment is a necessary — and possibly a sufficient — condition for that result), whereas to say that the norm-subject is permitted to choose between different courses of action — or that, within one course of action, at a certain point he can choose between different subcourses of action — means that to follow some of those courses or subcourses of action is a necessary — and possibly sufficient — condition for the result. 'Obligatory' and 'permitted', here, have an anankastic (or, from the point of view of their users, a technical), rather than a deontic meaning: in the first case, they indicate that the norm-subject, in order to bring about the result, *must* follow a certain course of action; in the second, that, in order to bring about the result, she *may*

— in the non-deontic sense — choose between different courses or subcourses of action, and *must* choose one of them.⁹

What has been said so far seems to enable us to confront, at least in part, the argument of the practical irrelevance of permissive norms given by Echave/Urquijo/Guibourg. In their funny little story, the conclusion is precisely that "things would stay exactly the same, as long as it did not occur to Toro Sentado to prohibit something". As we have already seen with respect to Alchourrón and Bulygin, on the one hand the permissions issued by Toro Sentado imply prohibitions for possible future lower-ranking authorities: they are prohibited by Toro Sentado to issue norms that prohibit hunting on Tuesdays and Thursdays. On the other hand, by the mere fact of having elected Toro Sentado as their leader, the *charrúa* actually have changed their normative universe, by introducing, in our view, a rule conferring power to Toro Sentado — the power to issue norms that are binding for the other members of the tribe. And that, obviously, does change the expectations of the tribespeople. Of course, what seems strange to us is that among the *charrúa* there isn't someone — say, a Hartín — sufficiently alert to realize that the election of Toro Sentado has changed things, and a few other circumstances as well: first of all, that — in contrast to the idea of an empty normative universe which, according to the example, the *charrúa* had of their own situation — that election presupposed at least the existence of a rule according to which the *charrúa* as a whole had the

---

⁹ In some of his papers, von Wright (1968 and 1969) proposed to construct deontic logic as a 'fragment' of the Logic of Conditions. Thus, for example, obligation is defined in terms of a necessary condition: 'Op' $=df$ '$Nc\,(p,\,I)$', that is: "To say that something ought to be, or ought to be done, is to state that the being or doing of this thing is a necessary condition (requirement) of something else" (1968, 5). Weak permission (equivalent to '$\sim O\sim$') also remains in the sphere of necessary conditions: it is the negation of the fact that the contradictory of some state of affairs (or some action) is a necessary condition for I. And strong permission is defined in terms of sufficient conditions: 'Pp' $=df$ '$Sc\,(p,\,I)$', that is, that p is permitted is equivalent to saying that p is a sufficient condition for some other thing. 'I' here is a propositional constant whose content does not need to be specified, although a typical meaning of it, in the sphere of legal norms, could be 'immunity from punishment'. Now, as González Lagier (1995) has correctly asserted, "the logic of norms based on the theory of conditionals is not really [...] a logic of deontic concepts, but rather a logic of the notions of *technical* ought and may". That means that such a logic cannot give account of deontic norms, but it may be applicable in the context of power-conferring rules. Specifically, we think that in this way one can account for the notions of 'permitted' and 'obligatory' with respect to element Y in our 'canonical form' of a power-conferring rule. If element Y is taken as a whole, one can say that action Y is a necessary (and, if supplemented with X, a sufficient) condition for result R (where R would be a 'translation' of I). But if one takes only a fragment of Y (say, action b, located between a and c), then to say that b is obligatory means that, given a, to perform b is a necessary and sufficient condition for c; whereas b is permitted if to perform b is a sufficient condition for c (and there is at least one other action d which, given a, also is a sufficient condition for c).

normative power to elect a leader (that is, to confer on someone the normative power to issue binding norms, and to impose or — as seems to be the case of the example — not impose on him obligations and prohibitions concerning the exercise of that power); secondly, that, if the power to issue binding norms is conferred on Toro Sentado, one must determine what exactly Toro Sentado must do in order to issue such norms or (what amounts to the same) what course, or courses, of action Toro Sentado must follow, so that his utterings must be recognized by the *charrúa* as constituting the issuing of a norm, or when an uttering of Toro Sentado must be regarded by them as constituting a reason for action and cannot merely be regarded as expressing an ought judgment by whose formulation Toro Sentado expresses the existence of reasons for action independent of his own utterings.[10]

2.3. Permission and principles. Constitutional freedoms

Besides permission referring to natural actions and to the exercise of normative powers, the law knows a third type of permissive sentences: constitutional freedoms which, in our view, require separate treatment. As we saw in the contexts of Ross, von Wright, and Alchourrón and Bulygin, it can be tempting to regard constitutional permissions as *equivalent* to prohibitions of interference addressed to the legislator and, in general, to lower-ranking authorities. But this reduction is not acceptable. It is, of course, true that prohibitions to the legislator and, in general, to lower-ranking authorities *are derived* from constitutional permissions; but the constitutional permissive sentence is located, so to speak, on a higher justificatory level. In order to justify these assertions, we

---

[10] On that point, in general, cf. Bayón 1991a, 248 ff. Because one and the same sentence — for example, 'One day a week, there should be no hunting' — can express either the issuing of a norm (a rule) or merely of an ought judgment. In the latter case, Toro Sentado would not be speaking as an authority, since he would have no other intention than expressing the existence of certain reasons for action, independently of the fact that he formulates that sentence: for example, that by abstaining from hunting one day a week the animal population can be kept in equilibrium, or that this allows one to dedicate one day a week to the maintenance of the tents, etc.; in the first case, that is, in the case of a norm being issued, Toro Sentado would say 'One day a week, there should be no hunting' with the intention that his uttering that sentence be regarded as a peremptory reason for action, that is, for doing what has been ordered, regardless of the judgment the addressees themselves may have about the reasons pro and con (although, of course, the issuing of the norm itself does not need to be arbitrary, and can itself be based on reasons). Only in that case, Toro Sentado acts as a 'leader', that is, as an authority. That is why it is necessary — as Bayón points out — "to have some *criterion of recognition* counting as an accepted signal" (ibid., 270) that can indicate when Toro Sentado issues a norm, and not merely an ought judgment.

will again use an example, consisting in two provisions of the Spanish Constitution:

1) "It is the task of the public powers to promote the conditions necessary for the liberty and equality of individuals and of the groups to which they belong to be real and effective; to remove the obstacles that prevent or hinder their full effectiveness; and to facilitate the participation of all citizens in political, economic, cultural and social life" (art. 9.2).

2) "The following rights are recognized and protected: *a)* Freely to express and diffuse ideas and opinions orally, in writing, or through any other means of reproduction [...] *d)* Freely to communicate or receive truthful information by any means of distribution" (art. 20.1).

As will easily be granted, the content of article 20.1 can be paraphrased (at least partially) as follows: 'It is permitted to express or not to express, and to diffuse or not to diffuse, any idea, opinion or truthful information'.

In ch. I, we argued that the special characteristic of legal principles — as opposed to rules — is the *open* configuration of their conditions of application. Accordingly, we now paraphrase article 20.1 in the following form: 'Unless in a specific (generic or individual) case there are applicable principles commanding otherwise which, in that case, have higher weight, it is permitted to express or not to express, and to diffuse or not to diffuse, any idea, opinion or truthful information.' Now, the question is whether this formulation (which we will call E_1) is *equivalent* to the following *(E_2)*: 'Unless in a specific (generic or individual) case there are applicable principles commanding otherwise which, in that case, have higher weight, it is prohibited to the legislator and, in general, to all public powers to impose obligations or prohibitions concerning behaviour consisting in expressing or not expressing, and diffusing or not diffusing, any idea, opinion or truthful information, to hinder that behaviour, or to impose sanctions as a consequence of such behaviour.' In our opinion, the two formulations are not equivalent, for the simple reason that the first *(E_1)* more directly expresses the value that can serve as a justificatory foundation for the second *(E_2)*, while the inverse is not true. It makes perfect sense to say that the Spanish Constitution regards the freedom of expression and diffusion of ideas, opinions, etc. as something valuable and therefore imposes on the public powers the prohibition to interfere with it (unless there are principles applicable and of higher weight in the respective case that command otherwise and thus justify the intervention). In contrast, it does not seem to make much sense to say that the Spanish Constitution regards interference by the public powers with the expression and diffusion of ideas, opinions, etc. as a disvalue and, therefore, permits their free expression and diffusion. This means that E_1 *presupposes* that there are reasons that justify the freedom of expression (and, therefore, also justify the prohibition of interfering with it). Thus, the answer to the question 'Why can I express my opinions freely, and why may nobody hinder me from doing so?' or 'Why may the legislator not issue a norm containing restrictions of that freedom?' cannot simply be to invoke E_1: it must also include a reference to a value judgment of the type 'I may freely express my opinions because freedom of expression in itself is something valuable, since it is an essential component of personal autonomy and political democracy'.

In Carlos Nino's categories,[11] a constitutional permission does not simply express a prescription — a directive —, but also a value judgment.[12] To this one could object that, if the difference between value judgments and prescriptions basically consists in the fact that issuing a value judgment means to express that there are reasons supporting it, then it seems to be almost a contradiction to speak of a value judgment issued by an act of authority. In our opinion, however, that objection can be countered if one considers the following: In the case of prescriptions, the act of an authority is of a constitutive character; there are prescriptions because there are acts of prescribing performed by authorities. Authorities, however, cannot create values but only recognize them.

Constitutional permissive sentences, besides, are 'translated', in the sphere of directive norms, not only into mandatory principles, but also into policies. This can be clearly seen if article 20.1 is linked to article 9.2. From their combination, one can derive a sentence *(E_3)* like the following: 'The public powers must adopt the appropriate measures so that all individuals, and the groups they belong to, have real and effective possibilities to express and diffuse their ideas, opinions, etc.' In our opinion, while E_2 expresses a *principle in the strict sense* — ordering public powers to abstain from any type of conduct that would constitute an interference with the freedom of expression, in all cases where that principle prevails over principles pointing in the opposite direction —, E_3 expresses a *policy* or *program norm*. In contrast to principles in the strict sense, these norms do not command or prohibit any particular type of conduct. They only indicate that it is obligatory to pursue certain ends, certain

---

[11] Cf. Nino (1985a).
[12] In ch. IV, we will undertake a general examination of the distinction between the directive and the evaluative aspect of norms. Here, when referring to mandatory principles and, later on, to policies, what we will basically consider will be the directive aspect.

states of affairs; but they leave it to the discretion of their addressees (for our purpose, to the discretion of the legislator and, in general, of the public powers) to choose the appropriate means. Thus, while the principle that prohibits interference with free expression constitutes a constraint for the objectives that can be pursued through the legislative and more generally through the political process, the policy commanding to pursue the end that individuals and groups have the real possibility to express themselves stipulates an objective the legislator and, in general, those who participate in the political process cannot (legitimately) not want, but it leaves open the question of what means are best suited to attain that goal.

3. Some conclusions

The fundamental conclusion to be drawn from all this is that the meaning of permission is different in each one of the three contexts we have distinguished, and that it is this variety of contexts of usage (rather than the distinction between weak and strong permission) that is decisive. Seen exclusively as operators of regulative rules, permissions can be translated in terms of the enactment and derogation of mandatory norms and, possibly, definitions. They express the absence of a peremptory reason (clarifying that there really is none, or cancelling one that exists) for some particular conduct. In the context of the exercise of normative powers, one must distinguish whether the operator 'permitted' modalizes the normative result or the action that is a condition for bringing about that result. In the first case, permissions have a regulative or deontic character and can be translated in terms of the derogation or the simple negation of mandatory norms, but their function is not only that of expressing that there are no operative reasons. In relation with the power-conferring rule, they also have the function of qualifying the hypothetical reason constituted by the power-conferring rule as problematic. In the second case (when 'permitted' refers to the course of action — the element we have called Y — in relation with the institutional result or normative change R), permission does not have a deontic or regulative character; rather, it expresses an anankastic possibility which, for its users, translates into a technical-institutional possibility. Finally, constitutional permissions are not only indirectly expressed directives, but also value judgments. From them, one can derive directives in the form of mandatory principles and policies, but they are not equivalent to them, because their justificatory scope is substantially more extensive.

APPENDIX TO CHAPTER III
A NOTE ON CONSTITUTIONAL PERMISSION AND BASIC RIGHTS

The characterization of constitutional permissive sentences we have just presented is, in our view, perfectly compatible with an adequately elaborated theory of basic rights. As an example of such a theory we will use the conception presented by Francisco Laporta (1987)[1] which we will briefly sketch.

In this paper, Laporta has two objectives: to clarify what it generally means 'to have a right'; and to analyze the structural and formal characteristics normally attributed to the notion of human right, that is, their universality, absoluteness and inalienability. Concerning the first, Laporta's thesis is that 'to have a right' cannot be paraphrased entirely in normative terms. Rights are *prior* to the *normative protection* granted to them. Rights are entitlements, i. e., reasons justifying the existence of certain norms, but not norms themselves:

"What I want to suggest", Laporta (1987, 27 f.) writes, "is that 'rights' are something, so to speak, *prior* to actions, claims or requirements, *prior* to normative powers, *prior* to normative freedoms, and *prior* to immunities of status. They are better understood if they are conceived as the *entitlement* [...] underlying all those, and other, techniques of protection, that is, if they are seen as what justifies setting all those techniques in motion. I suggest that when we use the notion of 'right' we are not referring to any primary or secondary norms of some normative system, but to the reason [...] that is presented as the justification for the existence of such norms."

From this characterization, Laporta derives three very important conclusions: *1)* rights are not exclusive to legal systems; it makes perfect sense to speak of moral rights; *2)* the core of the notion of a right is *something* prior to norms, but existing in legal systems (which not only consist of norms, but also of definitions, descriptions of states of affairs, and value judgments); and *3)* normative systems are not only deductive systems; among their elements, their are also — non-deductive — relations of justification or of an instrumental kind.

---

[1] In the same number of the journal in which Laporta's paper is published, several critiques of it (our's among them) plus a reply by Laporta also appeared. In the reply, he says that "[Atienza and Ruiz Manero] do not seem to criticize the main thesis, namely: that the meaning of 'having a right' cannot be exhaustively accounted for through typical deontic expressions". In fact, we do not; but until now we had not realized to what extent that is true. This means that we agree with Laporta's thesis to a much higher degree than what we had thought. A very critical view of Laporta's theses can be found in Vernengo (1990).

Concerning the second objective — to clarify the universality, absoluteness and inalienability normally attributed to human rights —, Laporta formulates three fundamental theses:

1) Universality does not refer simply to a formal logical predicate; it says something substantial about the subjects of those rights; the characteristic of 'universality' means that human rights are ascribed to *all* human beings:

"While purely logical universality permits one to put any kind of circumstance of the case, condition of the subject, or characteristic of the context into the universal sentence (for example: 'For *all X* such that *X* is in circumstance *A*, satisfies condition *B* and lives in context *C*, *X* has the right to ...'), the universality of human rights requires precisely that one disregard such circumstances, conditions and contexts, because those rights claim to be ascribed to *everyone*, unconditionally. Apparently, it is enough to fulfil the minimal condition of being a 'human being' in order to have those rights ascribed to one (for all *X* such that *X* is a 'human being' — irrespective of the context and circumstances —, *X* has the right to ...')" (Laporta 1987, 32).

According to Laporta, this characteristic places human rights into the sphere of ethics.

2) As for their absoluteness, human rights appear as moral claims endowed with a characteristic force deriving from the fact that "they are the expression of goods of special relevance for human beings" (ibid., 37); they are "the strongest moral requirements to be had in moral discourse" (ibid., 41). This means that *a)* they cannot be the object of trade-offs or negotiation, i. e., they always prevail over other claims — even moral ones — not entailing rights; and *b)* they can only be displaced by conflicting moral claims of equal value, i. e., by other human rights (in that sense, their absoluteness is *prima facie*).

3) Finally, their inalienability means that "human rights cannot be renounced even by their own holders" (ibid., 43):

"Just as *everyone* has the obligation to respect the right of everyone else, and cannot change that right, the holders of a right themselves have the obligation to respect their own rights, that is, they are normatively immunized against themselves." (Ibid., 44)

Now, if one compares this conception of rights with our reconstruction of constitutional freedoms (which are a subclass of the class of human or basic rights, grounded on the value of autonomy), one easily sees that they fully coincide.

Laporta's idea that rights are prior to norms we had expressed by saying that constitutional permissive sentences contain an aspect of a value judgment that has priority over the directive aspect of the prohibition to interfere. Also, the consequences Laporta draws from his general thesis have (in our conception of constitutional permissions) the following three correlates:

ad 1): That rights are not exclusive to legal systems translates into the idea that the value judgment expressed in the permissive sentence is not created by the authority; authorities cannot create values — basic rights — but only recognize them.

ad 2): The entities Laporta locates prior to norms and which are the core of the notion of a right are, in our conception, value judgments; what happens is that, as will be seen in the next chapter, we understand that value judgment as something internal to the norm; more precisely, we think that in norms two aspects must be distinguished: an evaluative aspect, and a directive aspect.

ad 3): Between E_1 and E_2, there is not a deductive logical relation, but a relation of justification: E_1 justifies E_2, but not *vice versa*.

As for the three characteristics of human rights, the correspondence with our conception is as follows:

Universality, in our opinion, is connected to the open configuration of the conditions of application that characterizes principles and, especially, the class of principles we are interested in now: those deriving from constitutional freedoms. The 'claim' of universality of rights — rights to freedom — implies that the antecedent of the corresponding normative (conditional) sentences contains very abstract properties. Nevertheless, the translation of such basic rights into legal terms does have an effect: on the one hand, the claim of universality can be — and usually is to some degree — restricted by peculiarities of the respective legal system, as, for example, their national character; on the other hand, since what justifies legal systems is that they make it possible for practical reason to go beyond what general practical discourse, i. e., moral discourse, would permit, the transformation of rights into fixed legal configurations is not confined to the level of principles, but is continued in mandatory rules, power-conferring rules, definitions, etc.

As for the absolute character of rights, this is exactly what made us emphasize the difference between policies and principles in the strict sense. From the fact that constitutional freedoms are expressed in that kind of principles, we can derive two consequences parallel to those indicated by Laporta: that there can be no trade-off or bargaining, in our model corresponds to the impossibility of submitting them to a process of optimization; and their *prima facie* character corresponds to the fact that one must take into consideration the claims deriving from other principles in the strict sense, in order to determine which should prevail in each case. This does, of course, not mean to deny that rights — freedoms — also translate into requirements that have the form of policies. But the essential idea is that the *core* of rights — of freedoms — is constituted by principles (and mandatory rules derived from principles).

Finally, inalienability, understood in the sense we have seen above, seems to conflict with the idea of constitutional freedoms or permissions, that is, with human rights based on the value of autonomy. But this incompatibility is only apparent. We can without any problem stick to the idea that constitutional freedoms are inalienable, in the sense that their holders cannot renounce them (one cannot renounce the *right* to life, or to freedom of expression), although they can renounce their exercise (one can renounce to go on living, or to express any kind of opinion). This is exactly how Laporta understands it too:

"[O]ne does not have the freedom to have or not to have basic rights, although some basic rights consist in having some freedom. Freedom, in this case, refers to the exercise of rights, not to having them, which is always imposed on the individual." (Laporta 1987, 44)

To conclude (and this is probably the most important lesson to be drawn from all that has been said here), sentences expressing constitutional freedoms cannot be seen exclusively in terms of directives; they must also be conceived in terms of value judgments. In the next chapter, we will extend this thesis to all normative sentences. We think that in this way we can contribute to filling a substantial gap, since, unfortunately, a theory of values — or of value judgments — in the law is still almost entirely missing.

CHAPTER IV
VALUES IN THE LAW

1. Introduction

In the previous chapters we have repeatedly mentioned values. For instance, when we examined principles we said that principles in the strict sense imply the assumption of "values regarded as categorical reasons with respect to any interest". Therefore, we said, norms transporting those values — i. e., principles in the strict sense — always prevail over policies and play a predominantly negative role: they prevent that the pursuit of some interest harms those values. Now, this does not mean — we added — that policies are not also supported by values, at least if that expression is used in a wide (and common) sense; rather, we can say that what is implied here is another kind of values to which we then referred as social objectives, collective interests of an economic, social, cultural, etc. character.

Also, when we analyzed the question of permission, we reached the conclusion that sentences expressing constitutional freedoms are not *equivalent* to the sum of prohibitions of interference and optimization mandates that can be derived from them; rather, they are located on a higher justificatory level. Such sentences also express, we said, the *value judgment* serving as the foundation for the corresponding principles and policies. Only if seen in this light, the justificatory role and the special expansive force constitutional freedoms have in legal discourse can be explained. From this, of course, it does not follow that that important role of value judgments is confined to provisions expressing constitutional freedoms and does not also cover those directly expressing mandatory principles. Our thesis, which was more or less implicit, and which we will try to explain and develop later, is that principles — and also, in different senses, the other norms of a legal order — not only contain a *directive* (or normative, in the strict sense), but also an *evaluative* element.

As will be seen in ch. V, this is also true with respect to the rule of recognition which, besides providing a theoretical criterion for the identification of legal norms, has two practical dimensions: as a guide for behaviour (especially the behaviour consisting in the making of legal decisions) and as a criterion for evaluation (again, especially of such decisions).

Now, in the earlier examples, the reference to value judgments is not contained in the sentences that make up a legal system; they appear when — from the perspective of legal theory — one either tries to elucidate or explicate their meaning (in the case of sentences expressing mandatory principles in the strict sense, policies or constitutional freedoms) or to explain the operation of the 'master norm' — the rule of recognition — that gives us the criterion by which to determine which sentences belong to some legal system. However, in the language of the legislator one can find examples of sentences explicitly and directly expressing value judgments. This is often the case — indeed, it is probably inevitable — when one looks at the "statement of purposes" of a law. And it is also not uncommon, although less frequent, in the articles of statutes. An interesting example of the latter is article 1.1 of the Spanish Constitution: "Spain is constituted as a social and democratic State under the rule of law, declaring as the highest values of its legal order liberty, justice, equality and political pluralism." Besides, reference to value judgments — or, simply, to values — abounds in legal language, especially when it comes to grounding a decision in those hard cases that are difficult because their solution requires a balancing of different values (for example, liberty and security; honour and freedom of expression; life and personal autonomy, etc.); and also — not surprisingly — in the language of legal dogmatics, especially those constructed over those parts of the legal order where the values regarded as the most basic ones of a society have the greatest impact. Thus, in the dogmatics of criminal law, for some time now there has been a dispute between two positions about the conception of criminal norms: as a imperatives, or as value judgments. We will look at this dispute somewhat more closely now.

2.   Two conceptions of criminal norms

As formulated by Mir Puig, the two opposed positions are the following:

"Those who conceive of a criminal norm as a 'norm of valuation' regard it as an expression of a *value judgment* distinguishing what is licit according to criminal law from what is illicit. In that sense, article 407 of the Criminal Code is nothing but a judgment about the disvalue of killing another person. Thus, it does not contain an *imperative*, addressed to citizens, not to kill.
The *imperative* theory, in contrast, regards a criminal norm as a norm of *determination*, that is, as a *command* or *prohibition* addressed to the citizens. In that sense, the legal order consists of expressions of the legislator's will requiring a certain behaviour from the members of the legal community, and its norms are obligatory prohibitions the people concerned must comply with."
(Mir 1976, 53 f.; also Mir 1990, 42 ff.)

Now, understood in a radical sense, both theses seem to be untenable. One cannot possibly deny that a criminal norm like article 407 implies more than just a value judgment, and that if someone is threatened with a sanction of up to twenty years in prison this must be so because his behaviour — killing another person — deserves a negative value judgment from the legislator. The difference between the two positions is, rather, that each of them gives priority to a different one of the two elements of a criminal norm: the imperative, and the evaluative. Thus, Mir Puig himself, for example, defends the imperativist conception of a criminal norm as follows:

"Criminal norms operate by appealing to the citizens' *motivation*, by threatening them with the harm of punishment, in order to incline them to decide in favor of the law and against crime [...] Obviously, the imperative is logically preceded by a negative valuation of the prohibited or commanded conduct, but that valuation is only an internal instance, within the legislative process, whereas what is decisive for the efficacy of the criminal norm is that the legislator provide it with the force of an imperative. That is what distinguishes a mere *wish* from a valid *norm*." (Mir 1976, 57; in the same sense also Mir 1990, 42 ff.)

As an example of the opposite position, which gives priority to the evaluative aspect, we can quote the treatise on criminal law by Cobo and Vives,[1] who — following Mezger — distinguish between the *objective norm* of valuation (stipulating the 'material wrong', the *unlawfulness* of an action) and the *subjective norm* of determination (determining whether or not the author of an action is *culpable*):

"'This determination [of what is and what is not in accordance with the legal order] is done according to the norms of the law which, therefore, appear as *objective norms of valuation*, as judgments over certain events and states from the point of view of the law.'[2] The norm appears, *prima facie*, as a judgment over reality, or rather, those parts of reality that acquire relevance for social life, whether they be human behaviour or simply natural events. That judgment qualifies those events, states or actions as positive or negative, according to the purposes of the law. To that conception corresponds, in the criminal order, the conception of material wrong *[Unrecht]* as an objective violation of the norms of valuation, as a harming or endangering of interests (linked to events, states, etc.) qualified by those norms as legal goods.
From those objective norms of valuation, 'the *subjective norms of determination* are deduced', that is, the directives for behaviour derived from them. The norm as an objective norm of valuation is a *norm of law*: it determines the order in which social life is embedded, and represents the

---

[1] In view of the 'germanization' predominant in Spanish criminal law, one can say that the two positions mirror those held in Germany, especially by Maurach and Jescheck, in the first case, and by Mezger, in the second.
[2] The quotations within the quote are passages from Mezger's work.

solution to the many conflicts arising in life in society, according to the ideas that informe the legal system. The norm of determination is a *duty norm* arising from the legal qualification of a situation or state of affairs and personally obliging the particular citizen. In this sense, the norm is relevant for deciding whether there is *culpability* in a behaviour, since the personal blame culpability consists in can only be grounded in the existence of an *obligation* of acting differently from how one acted." (Cobo/Vives 1990, 213)

On the other hand, where imperativists think that criminal norms have a function of 'motivation', defenders of the theory of the 'double function of the criminal norm' see things differently:

"It seems more in accordance with the facts to say that *the main function of rules is not to 'motivate', but to produce distributions of goods and values between the members of the community, to protect the resulting distribution, and only in a secondary way to motivate individuals to respect it.* Thus, when a norm is regarded as a social fact, one can also attribute a double function to it: an objective one of distribution and protection, and a subjective one of motivation." (Cobo/Vives 1990, 214)

The supporters of both positions think that each one of them has important consequences from the systemic point of view; for example, imperativists see the subjective moment of disobedience, the intention, as belonging to the core of the material 'wrong', of unlawfulness; in contrast, supporters of the evaluative thesis — or of the thesis of the double function — defend an 'objective' system, that is, the separation of the objective 'wrong' (unlawfulness) and its subjective blameworthiness (culpability) (Vives 1979, 211). And this is said to apply also from the practical point of view, since a number of problems related to the relevance of error, to negligence, to participation, to failed attempts, etc. are said to be solved differently by the two positions. Here, the dispute supposedly is about the extent to which criminal law may legitimately intervene in the life of individuals. This last claim seems — to say the least — doubtful to us, because the basic assertions of each theory are later nuanced with a number of other postulates, which makes it impossible to say *a priori* that to adopt one or the other *necessarily* forces one to sustain a certain practical solution.[3] We will not here go into these questions which belong to the specific realm of criminal law, but we do want to explain why, in our opinion, both constructions are theoretically inadequate. Both of them do, of course, enable one to come to a 'reasonable' solution of the practical problems mentioned above, but only with the help

---

[3] See, for example, Silva (1992, 340 ff.) who shows how, from an imperativist conception one can avoid all — or almost all — the 'negative consequences' attributed to that position by critics.

of a number of tricks (which is probably higher in the case of the imperativist conception).

We think that mainly two objections can be directed against an imperativist conception like that of Mir. The first concerns the scope of his thesis. Because Mir constructs the theory in order to account for 'primary' criminal norms, that is, those addressed to the citizens. Thus, he doesn't deny that in the legal order there are other kinds of sentences — or norms — that cannot simply be seen as imperatives.[4] Also, within the sphere of criminal norms, he distinguishes between the already mentioned 'primary' norms, 'secondary' norms (addressed to the judge, ordering him to inflict some punishment), and norms "related to security measures".[5] In his opinion, it is clear that secondary norms cannot "have but an imperative nature" (Mir 1990, 43); that means that here one cannot even say that there is a value element. Because it would certainly not be comprehensible why a negative value judgment of the corresponding action should be underlying the imperative addressed to citizens ('It is prohibited to kill') while the imperative addressed to judges ('If someone has killed another person, it is obligatory to impose a minor prison term') would be a pure command which we would not need to suppose to be supported by any underlying value judgment. Obviously, one cannot argue that the imposition of prohibitions on citizens needs to be justified in terms of value judgments, whereas

---

[4] Cf. Mir 1976, 54, where, for example, he refers to power-conferring norms.

[5] We will not go into that last kind here; but it may be interesting to call attention to one point of their treatment by Mir. In his opinion, a norm stipulating security measures, "in contrast to criminal provisions, cannot be a command or prohibition addressed to citizens, nor a legal-ethical valuation referred to their addressees. They cannot direct an imperative at citizens, because they do not refer to *behaviour* that is prohibited, but to states of affairs in which a subject is dangerous, and one cannot imagine that it could be *prohibited to be dangerous*. Imperatives cannot refer to a way of being, but only to a way of acting. But one also cannot think that bills foreseeing security measures presuppose a legal-ethical valuation, because the latter can only concern conducts a subject can avoid, and not personal characteristics like being dangerous. What happens is that the provisions stipulating security measures do not entail any 'primary norm' addressed to citizens: neither an imperative norm, nor a norm of legal-ethical valuation. They only contain the norm addressed to the judge ordering him to impose a security measure on dangerous subjects" (1990, 45). We think that what Mir says here is not objectionable, provided it is understood as referring to the *concept* of a norm related to security measures. But that should not make us forget that under the label of a 'norm related to security measures' one can accomodate — as has frequently been the case in history — provisions ordering the judge to impose certain kinds of unpleasant treatment as a consequence of certain activities or conducts, such as 'habitual vagrancy and begging'. In those cases, it seems obvious that the legal order attempts to deter from performing those activities or conducts, and that, in this sense, it makes perfect sense to think that a provision of that type contains a prohibition of habitual vagrancy or begging, and a negative 'legal-ethical valuation' of them. That such cases constitute instances of a 'perverse use' of the institution of security measures is, we think, uncontrovertible; but that does not affect our point.

no such thing is required when it comes to impose the obligation on judges to impose sanctions on those same citizens (in case they violate those prohibitions). It seems to us that the need for justifying the imperative is even more obvious (and the justification itself more problematic) in the second than in the first case; because the imperative 'It is prohibited to kill' *commands to abstain from performing a disvaluable action*, whereas the imperative 'It is obligatory to inflict such-and-such punishment' *commands to do something* (deprive a subject of his freedom for a certain period of time) which is, in itself, *disvaluable*. Besides, if this were not so, then the entire discourse — and controversy — on the justification of punishment would literally be 'non-sense'.

The second critique refers to the idea that valuation is only an instance *preceding* the formulation of the imperative, that is, a kind of causal antecedent whose function ends as soon as one passes from the level of a 'wish' to that of a 'norm'. One of the consequences of this thesis is that apparently it cannot adequately account for the treatment criminal codes give to persons considered legally not responsible for their actions. Normally, it is said that children and the mentally impaired can perform unlawful actions, but cannot be culpable (they cannot be blamed for their behaviour). According to the imperativists, at least in the case of subjects who cannot in any way be motivated by norms (small children, or people suffering from severe mental disorders), one would have to conclude that their behaviour cannot even be unlawful (Silva 1992, 346 ff.). But that means that they give up a distinction which certainly seems to be important. When there are reasons of justification (for example, legitimate defence), it seems to make sense to say that, according to the legal order, the behaviour is permitted, that is, the complete norm would say that it is prohibited to kill *unless* (among other possible circumstances) there is a situation of legitimate defence, in which case the prohibition is lifted, that is, the behaviour in question becomes permitted. But that is not true with respect to reasons for the exemption from culpability; it makes no sense to say that a (complete) criminal norm permits those not responsible for their actions (for example, small children, or the severely retarded) to kill. But, then, if criminal norms are nothing but imperatives, what is the deontic status of the actions of persons who are not responsible for them? Are they simply indifferent from the point of view of criminal law? Does it make sense to say that the killing of someone by a severely retarded person or a small child is the same as the killing of a fly, to use Welzel's famous example? To this, an imperativist could possibly object that the behaviour of a person not responsible for her actions conforms to the material facts of a case [i. e., to what in German jurisprudence is called a *'Tatbestand'*] (and, therefore, is not indifferent), but is not unlawful. We think, however, that this means to abandon imperativism, insofar as it implies admitting that a criminal norm does something more than merely qualifying a behaviour as obligatory, prohibited, permitted, or indifferent.

Criticism must also be voiced concerning the conception of the double function of a criminal norm, as presented in the work of Cobo and Vives. The critical points probably derive, more than anything else, from a certain lack of clarity on the part of the authors. As we have already seen, this conception has as a consequence — or claims to justify — an 'objective' systematic construction "in which the sphere of disvalue (of the object of protection) corresponds to *unlawfulness* and the imperative or motivational aspect of norms to *culpability*" (Cobo/Vives 1990, 214). Now, this precise distinction between the level of disvalue — said to be the only one corresponding to unlawfulness — and the motivational or imperative level — said to be the only one corresponding to culpability — seems untenable. Because, in our view, for reasons we will see below, to say that 'X is disvaluable' implies that 'X ought not to be done'. Thus, the element of unlawfulness contains an evaluative aspect as well as a directive or normative (in the strict sense) one, although the former may have more weight. What happens is that the 'directive' message contained in, or implied by, the judgment of disvalue is a message that says nothing about the 'scope' of that judgment with respect to its addressees, which is determined together with the question of culpability: the reasons for exemption from culpability negatively delimit the sphere of the addressees of the directive part of a criminal norm. But that does not imply that culpability must be linked exclusively to the directive (or, if you prefer, imperative or motivational) part of a criminal norm, disregarding any evaluative element. Because the reasons of exemption from culpability are not only based on the fact that there are certain categories of subjects who cannot be motivated by norms, but also on the fact that we would evaluate negatively (i. e., we would not think it justified) to inflict punishment on a subject with the characteristics constitutive of those categories.

Actually, in our opinion, the followers of the theory of the double function of criminal norms are right when they hold that a criminal norm has a directive or imperative as well as an evaluative element. They are also right when they postulate a relation of priority — *justificatory* priority, we would add — of the evaluative element over the directive one. But they are mistaken in dividing the two elements the way they do.

3. The double-faced character of norms and value judgments

The controversy among criminalists is, in our view, a good starting point for the study of legal values. Let us, therefore, take a closer look at some of the questions that have arisen in that dispute.

We said that a reconstruction of the circumstances exempting from culpability in terms of permissions would not make sense, in contrast to what is the case with reasons of justification, that is, those circumstances that exclude unlawfulness. We also said that one is permitted to kill in legitimate defence, but that an insane person is not permitted to kill, although she cannot be blamed for her behaviour. In other words, the behaviour of the insane brings about a result that is a disvalue from the point of view of criminal law; but an insane person cannot be guided by norms, and that makes it impossible for us to blame her for her behaviour and, thus, to be justified in punishing her.

It is that incapacity of being guided by norms which excludes the insane from the addressees of the directive part of criminal norms. The situation we confront when there is a reason for exemption from culpability is a situation where, for a certain category of subjects (those to whom the respective reason for the exemption from culpability applies), the judgment of disvalue concerning a certain behaviour does not translate into a prohibitory directive (nor any other kind of directive) of that behaviour.[6]

Earlier, with respect to constitutional liberties, we saw another case where there was no complete correspondence between a value judgment and a directive: That type of permissive sentences — we said — has directive consequences, that is, they translate into mandatory principles and policies, but they are not equivalent to such directives, because they are also value judg-

---

[6] We can regard the relationship between material facts, unlawfulness, and culpability as a chain of *prima facie* judgments of value and duty that are transformed into judgments of value and duty *all things considered* (that is, definitive judgments). Thus, the element of the material facts tells us that a certain action is *prima facie* disvaluable (or, 'transposed' to the normative level, that *prima facie* it should not be performed). The element of unlawfulness says that some action *a)* is disvaluable all things considered (should not be performed, all things considered) and *prima facie* blameworthy (and that, therefore, its author should *prima facie* be punished for it), unless there is a reason of justification, and *b)* is legitimate all things considered (is permitted, all things considered) and cannot be blamed on its author, if there is a reason of justification. And the element of culpability says that the action is *a)* blameworthy all things considered (and, therefore, the author should be punished, all things considered), unless there is a reason for exemption from culpability, and *b)* not to be blamed, all things considered, on its author (who, therefore, should not be punished, all things considered) if there is a reason for exemption from culpability.

ments whose domain — whose expansive force — is greater than that of the directives deriving from them.[7]

---

[7] With good reason, Juan Carlos Bayón has pointed out to us another case of incomplete correspondence, located outside of the sphere of law. It is the case of acts known as 'supererogatory' in moral theory. An act is called supererogatory if its performance is praiseworthy, but its forbearance is not blameworthy (because it violates no duty). A typical example is that of the soldier who throws himself on an exploding grenade and dies, in order to save his comrades' lives. Intuitively, everyone would probably accept that the soldier in question has done something that deserves a highly positive moral evaluation; but we would not say that because they did not do the same thing the other comrades failed to comply with a duty. Other examples of supererogatoy acts are less pathetic: if N, a very busy person, manages to visit her sick friend X every night, taking care every day to offer him entertainment and diversion in the form of books, videos, etc., we would say that N has acted better than if she had only visited X three times a week; but in the latter case we would not say that N had failed to comply with her duties as a friend. Thus, we can say of a supererogatory act that doing it is valuable, but not obligatory. Therefore, the evaluative and the directive spheres, to some extent, seem to separate.

The status of supererogatory acts poses very difficult questions for ethics because it is not easy to accomodate them within a consistent conception of morality: How is it possible that performance of an action which, in some context, is the most valuable of all possible actions, is not obligatory? Must we not always do what is best? Actually, some very influential moral theories leave no room at all for the concept of supererogatory acts: According to classical utilitarianism, for example, agents must always choose the course of action that produces the greatest global good, irrespective of the cost this implies for the agent himself. And if James S. Fishkin (1986) is right in categorizing them under the label of 'systematic impartial consequentialism (SIC)', the same would be true of moral theories as distant from utilitarianism as those of John Rawls or Bruce Ackerman: the repeated admission of supererogatory acts in *A Theory of Justice* would be inconsistent with Rawls' basic position — grounded on the equal consideration of everyone's interests — that agents may not give special weight to their own interests or those of persons close to them.

One way open to this kind of conceptions of morality for accomodating supererogatory acts is the one suggested (and rejected) by Raz (1986, 197 f.): supererogatory acts could be understood as acts that are in fact obligatory, but compliance with which requires such extraordinary personal qualities that, while their omission cannot be justified, it can, however, be excused. But it would hardly be accepted as an adequate characterization of what the soldier in the first example did to say that he complied with his duty, whereas his comrades failed to do so, although they had an excuse for it. Except for fanatics, everyone would agree that what the soldier in question did was beyond the 'limits of obligation' (for this expression, cf. Fishkin 1986 and 1982), and that such conduct could not be required, neither of him nor of his comrades.

This kind of considerations has led Bayón (1991a, 364) to reject the assertion sustained — he says — by many moral philosophers that the sentences 'There is a moral reason for performing acts of class P' and 'The acts of class P are *prima facie* obligatory' express exactly the same thing. In Bayón's view, with respect to the class of supererogatory acts, the first sentence is true — because "otherwise one would not understand why their performance is morally praiseworthy, and not morally indifferent" —, but the second is false, because of supererogatory acts "we don't say that they are 'obligatory', not even *prima facie*".

However, in our opinion there is a way that allows us to assert that supererogatory acts are *prima facie* obligatory as well as that they are acts the performance of which, all things considered, is beyond the limits of obligation. Let us return to the example of the soldier: If we ask what act it was the soldier performed, and we answer 'saving the lives of his comrades', then it was undoubtedly a *prima facie* obligatory act, since 'saving the lives of others' obviously is a *prima facie* obligatory act: everyone would agree that saving the lives of others under circumstances that do not imply a substantial sacrifice for the agent is obligatory, all things considered. What makes the

In order adequately to understand this kind of situations (which also perhaps helps us clarify the question of the nature of criminal norms), we think it can be very useful to return to the consideration of legal norms as reasons for action. Because when one says that a norm is a reason for action, one actually says two different things: that a norm is a guide for behaviour, and also that it is a criterion for the evaluation (that is, for the justification or critique) of behaviour. These two elements normally overlap, and that is why their duality usually goes unnoticed — or rather, in most cases, does not need to be noticed. In a way, they are like the two sides of one and the same reality. But — as we have seen — there are occasions when the two sides seem to separate. Thus, we could say that two aspects can be distinguished in a norm: the directive, or — if you prefer — normative (in the strict sense) one, that is, the one that guides behaviour; and the evaluative one that contains a criterion of evaluation (that is, of justification or critique).

Interestingly, that very same duality has also been said to characterize values.

---

act performed by the soldier supererogatory is that, in this case, saving the lives of others did imply a very substantial sacrifice on the part of the agent. To see things in this way implies that supererogatory (individual) acts are cases of *prima facie* obligatory (generic) acts, performed under circumstances — as in the case of the soldier — or through activities — as in the case of the daily visit to the sick friend — such that performing the action goes beyond the limits of obligation (on the distinction between *act* or *action* and *activity*, which comes from von Wright, cf. Appendix to ch II, n. 2). In this light, we think that Rawls is right when he says (1971, 439) that "supererogatory actions are ones that would be duties were not certain exempting conditions fulfilled which make allowance for reasonable self-interest".

If we go back once more to the example of the soldier, the condition for reconstructing things in this way is, of course, that the act carried out by the soldier is described as 'an act of saving the lives of others'. If we would describe it as 'an act of sacrificing one's own life in order to save the lives of others', the supererogatory would be located not in the circumstances of performance of the individual act, but in the generic act itself. The relevant difference between the first — 'an act of saving the lives of others' — and the second description — 'an act of sacrificing one's own life in order to save the lives of others' — is that the first, but not the second points to the consequence (the change in the world) constituting the agent's intention. And it is justified to point to the consequence constituting the intention of the agent — as a criterion for the description of an action —, because it is the intention that makes the act morally valuable: what the soldier did was morally valuable because it was done in order to save the lives of others and not, for example, in order to commit suicide or to be admired posthumously.

On the other hand, we agree with Bayón that the reason why supererogatory behaviour cannot be required is the moral relevance of the agent's autonomy, that is, the weight an agent may legitimately give to his own personal projects: the value of personal autonomy is the barrier which in the case of supererogatory acts prevents complete correspondence between the evaluative and the directive spheres, because in the contexts here described it does not allow one to pass from the assertion that 'X is the most valuable action, all things considered' to the assertion that 'X is obligatory, all things considered'.

According to Nicholas Rescher, to subscribe to a value is manifested in two different ways:

"First, on the side of *talk* (or thought), in claiming that N subscribes to a value, we give grounds for expecting a certain characteristic type of verbal action, namely, that he would 'appeal to this value', both in the support or justification of his own (or other people's) actions and in urging upon others the adoption of actions, courses of action, and policies for acting. Moreover, in addition to such overt verbal behavior we would of course expect him to take the value into proper account in the 'inner discourse' (*in foro interno*) of deliberation and decision making. In imputing a value to someone, we underwrite the expectation that its espousal will manifest itself, in appropriate ways, in his reflections regarding the justification and recommendation of actions. The prime indicators of value subscription are those items which reflect the *rationalization* (defense, recommendation, justification, critique) of aspects of a 'way of life'.
But second, on the other hand, we also expect the value to manifest itself on the side of *overt action*. We would draw back from saying that 'patriotism' ('financial security', 'the advancement of learning') is one of N's values unless he behaves in action — and not just at the verbal level — so as to implement the holding of this value by 'acting in accordance with it' himself, by endeavoring to promote its adoption by others, etc. In saying that prudence, for example, is one of N's values, we *underwrite the presumption* — this does not, of course, *guarantee* — that N behaves prudently (is a prudent person) [...] A value is thus bound up with a Janus-headed disposition cluster, and we expect it to orient itself in two directions, that of discourse and that of overt action [...]
Subscription to a value is consequently a two-sided affair, and value imputations have a double aspect: both verbal and behavioral. When we impute to the person N subscription to the value x, we underwrite the grounds for expecting from N a reasonable degree of conformity with the characteristic manifestation patterns of subscription to x both in discourse and in action. We thus impute a value to someone to characterize his vision of 'the good life' or at any rate his vision of how life ought appropriately to be lived" (Rescher 1969, 2 f.).

A similar idea, in the sense that acceptance of a value judgment cannot be separated from acceptance of standards of behaviour implied by that value judgment, can be found — referring especially to moral value judgments — in the work of R. M. Hare:

"[...] a moral judgement has to be such that if a person assents to it, he must assent to some imperative sentence derivable from it; in other words, if a person does not assent to some such imperative sentence, that is knock-down evidence that he does not assent to the moral judgement in an evaluative sense — though of course he may assent to it in some other sense [...] But to say this is to say that if he professes to assent to the moral judgement, but does not assent to the imperative, he must have misunderstood the moral judgement (by taking it to be non-evaluative, though the speaker intended it to be evaluative). We are therefore clearly entitled to say that the moral judgement entails the imperative; for to say that one judgement entails another is simply to say that you cannot assent to the first and dissent from the second unless you have misunderstood one or the other; and this 'cannot' is a logical 'cannot' — if someone assents to the first and not to the second, this is in itself a sufficient criterion for saying that he has misunderstood the meaning of one or the other. Thus to say that moral judgements guide actions, and to say that they entail imperatives, comes to much the same thing." (Hare 1952, 171 f.).[8]

Carlos Nino, however, has criticized Hare's view of the relationship between value judgments and directives for behaviour (especially when these directives — as is the case with most legal norms — originate in prescriptive acts by an authority). Nino stresses that, once the distinctive characteristics of evaluative judgments have been explained "in terms of reasons, one can clearly see how they differ radically [...] from commands and legal norms" (Nino 1985b, 116). The essential difference is said to be the following: Evaluative judgments indicate the existence of a reason for acting other than the formulation of the judgment itself, whereas imperatives, of which commands and legal norms are typical examples, "do not indicate the existence of reasons for acting; rather, they are assumed to constitute such reasons themselves" (ibid., 117).

In our view, however, Nino's disagreement with authors like Rescher or (explicitly) Hare rests on a misunderstanding provoked by the ambiguity of expressions like 'imperatives' or 'legal norms'. Here, we will concentrate on the case of legal norms. The following two things — among many others — can be understood by such a norm: on the one hand, a prescription itself, as issued by an authority; on the other, the ought judgment reproduced by the content of such a prescription. Consider the following argument:

1)   What A commands ought to be done.
2) A has commanded: 'In circumstances C, X should be done'.
3)   In circumstances C, X ought to be done.

---

[8] An explanation may be useful for a better understanding of Hare's conception. In the text, Hare speaks of 'evaluative moral judgements' and 'non-evaluative moral judgements' (or, as he calls them elsewhere, 'inverted-commas moral judgements'). With this somewhat strange terminology — 'non-evaluative moral judgements', 'inverted-commas moral judgements' — Hare refers to sentences concerning either the standards accepted as moral in a certain period or social group, etc. (for example, 'In Spain, in the forties, it was immoral for girls not to be virgins when they got married'), or the moral feelings of some individual or set of individuals (for example, 'The socialist representatives' consciences told them that they should vote against their colleague's petition'). Obviously, sentences like those contained in these examples do not express genuine moral value judgments: as Hare himself says, they are rather statements "of sociological fact" in the former and "of psychological fact" in the latter case. That consent with a moral judgment logically implies consent with the standard(s) of behaviour or, as Hare says, the imperative(s) deriving from it applies, of course, only to genuine (or, in Hare's terminology, to evaluative, not inverted-commas) moral judgments.

Obviously, if by a 'legal norm' (or 'imperative') we understand the prescription contained in *2)* (or the fact that that prescriptive act has been performed), Nino is right. And the fact that he identifies them with 'commands' indicates that Nino is, indeed, speaking of 'legal norms' precisely in that sense. But it is not in the sense of *2)* that we speak of 'legal norms' when we regard them as operative reasons or when we say that in a 'legal syllogism', the major premise is constituted by a norm.[9] In this context, the relevant sense of 'legal norm' (or 'imperative', if we want to stick with Hare's terminology) is that expressed in sentences like *3)*. Such sentences are ought-judgments deriving *a)* from the recognition of some person (or organ, or procedure, etc.) as an authority — which is expressed in sentences like *1)* —, and *b)* from the fact that that person (or organ, or procedure, etc.) has issued a prescription. And ought-judgments like that contained in *3)* have the 'double-faced features' of value judgments underscored by Rescher and Hare; that is, they operate as guides of behaviour as well as criteria for the evaluation of behaviour.

Now, if all that is right, we must conclude that the difference between value judgments and norms is one of degree, or emphasis. The question then is how the two aspects — the evaluative and the directive — can be understood and combined. Let us look at the following sentences:

1)   'Life is a good.'
2)   'Thou shalt not kill.'
3)   'It is prohibited to kill, unless there is a justificatory reason.'
4)   'Judges must sentence those who have killed another person to a minor prison term, unless there has been some justificatory reason or the author is a subject not responsible for his actions.'

---

[9] This can help us clarify why Nino insists that legal norms offer only auxiliary reasons for acting, and not operative reasons (cf. Nino 1985b). When formulating that thesis, Nino is thinking of a legal norm in the sense exemplified in *2)*, or one similar to it — a norm as an act of prescription or as a social practice — and not as an ought-judgment — of the kind exemplified in *3)* — which takes into account (as auxiliary reasons) the fact that certain prescriptions were issued by certain persons (organs, procedures, etc.), or the fact that there is a certain social practice. In any case, it is remarkable that an author like Nino, who probably has insisted more than anybody else on the plurality of concepts of 'legal norm' (cf. not only the papers contained in Nino 1985, but also Nino 1992 and Nino 1994), would enter into a dispute based merely on a misunderstanding.

Provisionally, we can say that *1)* is a sentence expressing a value judgment; *2)* is a statement of a principle (a principle norm); and *3)* and *4)* are mandatory rules with different addressees: those of *3)* are citizens in general, and those of *4)* are judges. But let us look at things more closely.

To say that *1)* is a value judgment is not, of course (in contrast to what Mir Puig seems to suggest), the same as expressing a wish. According to Rescher, we would say that for a person who accepts it, this value judgment implies that, *in principle* (let us assume that the assertion is not that life is an — or *the* — *absolute* value), he regards actions and omissions intended to maintain people's lives (again, let us assume the statement refers *exclusively* to the life of persons) as justified, and those provoking death as unjustified, and that he himself is disposed to perform or expects others to perform (also, in principle) actions and omissions intended to save life.

Now, *2)* does not seem to be very different from *1)*. To accept *2)* as a principle of behaviour also means to be disposed to preserve the lives of others, and that actions consistent with this principle would be considered justified and those in conflict with it unjustified and blameworthy. The two sentences stress different aspects, but other than that they say the same thing. Besides, in *1)* as well as in *2)* the axiological side has, so to speak, priority over the directive one: it makes sense to say that one should not kill because life is a good, or a value, but not that life is a value because one should not kill.[10] If we accept the convention that in a value judgment [like *1)*] there is special emphasis on the value element, and in a norm — a principle — [like *2)*], on the directive element, then we can say that *1)* — value judgment — has justificatory priority over *2)*, i. e. over principles.[11]

---

[10] The position sketched here is, of course, the exact opposite of the one defended by Kelsen; for him, in fact, norms — understood as directives that are the meaning of an act of will — constitute the foundation of all value judgments (Kelsen 1960, 18), and therefore it holds that: behaviour that is in accordance with a norm has positive, and behaviour that violates a norm has negative value (Kelsen 1960, 17).

[11] In our opinion, thus, whether a sentence has the form of a value judgment or of a principle does not make any difference for its meaning, but only for the relative emphasis of its evaluative or directive dimension. For example, article 1.1 of the Spanish Constitution now says: "Spain is constituted as a social and democratic State under the rule of law, declaring as the highest *values* of its legal order liberty, justice, equality and political pluralism." But the Constitution would not be any different (except for the substitution of 'political pluralism' by 'peace') if one of the other proposals had prevailed, which was: "Spain is constituted as a social and democratic State under the rule of law, declaring as the *principles* of its legal order liberty, justice, equality and peace." Another matter is that the language of values is usually more abstract than that of principles. For example, in the Spanish Constitution, the value 'equality' corresponds at least to the following different principles: equality before the law (art. 14); real or effective equality (art. 9.3); political equality (art. 23); fiscal equality ... But, then, the difference is not between evaluative sentences

3) and *4)* are clearly rules. Although in the conditions of application as well as in the legal consequences there are terms that may be vague, the configuration is of what we have earlier called 'closed form'. Thus, for those who accept those norms, *3)* and *4)* entail protected or peremptory reasons — and *3)*, besides, with respect to the reasons of justification, the absence of a peremptory reason — for doing what is stipulated, and for regarding behaviour that is in accordance with those norms as justified.

From *1)* to *3)* and *4)*, there is a process of specification; but in *1)* as well as in *3)* and *4)*, the two aspects of guiding and justifying behaviour are both present. The difference is that the guidance and justification offered by *3)* and *4)* is more precisely defined than in the other two cases, and that that guidance and justification is peremptory.

Now, the examples can be generalized in the sense that the two elements are not only present in (primary or secondary) criminal norms, but in all the kinds of norms we have distinguished: not only in mandatory rules (including — as we will see later — the rule of recognition) and in principles in the strict sense, but also in policies and in power-conferring norms. For example, a policy stipulating the goal of full employment is, *prima facie*, a guide for the behaviour of the public powers which, at the same time, serves as a *(prima facie)* justification of actions that have been or could be undertaken in that direction. In the case of power-conferring norms, the guidance and justification is indirect or hypothetical: if it is justified to bring about result R (either because a

---

and normative sentences of principle, but between sentences of a higher or lesser degree of abstraction; also, there would be nothing strange about speaking of the value of equality before the law, of the value of real equality, etc.

However, this is not the opinion of Gregorio Peces-Barba. In his view, the use of the term 'principles' in that article, instead of 'highest values', would have meant "to relapse into positions of natural law" (Peces-Barba 1984, 51), because that "would have suggested that there are *a priori* concepts which positive law must guarantee" (ibid., 52). We do not see how the opposition between natural law theory and legal positivism has anything to do with the use of the terms 'principles' or 'values' (Peces-Barba thinks that one of the main motives for having used the expression 'highest values' is that of overcoming that opposition [ibid., 53]); we do think, however, that his basic conception of the highest values incorporated in the Constitution, and the distinction he draws between them and the principles appearing in other constitutional articles is more or less correct. According to Peces-Barba, the difference lies in the perspective of the globality or totality of values, on the one hand, and in the fact that values have a normative nature, but are, in some sense, 'prior' to — that is, serve as the foundation for — the remaining norms (including principles), on the other (ibid., 39 ff.). In our terminology, this means that value judgments have a directive dimension, i. e., that they somehow guide behaviour, that the justificatory or evaluative element is prior to the directive one, and that the 'highest values' contained in article 1.1 are the most abstract, such that constitutional 'principles' can be seen as specifications of these values. But we do not think that any of this had changed if article 1.1. had used the term 'principle' instead of 'value'. In the same direction, cf. Rubio Llorente (1995).

mandatory norm stipulates that bringing it about is obligatory, or because bringing it about corresponds to a wish or interest of the agent and is not subject to normative constraints), given circumstances X, Y 'should' be performed. Here too, it makes sense to say that — given the circumstances — Y is a justified action, but here we have another kind of 'justification' than before. The reason is that the values implied in the different types of norms are not of the same kind. So we should examine this somewhat closer.

4. Types of norms and types of values

As will have been noticed, we are not attempting to construct a legal axiology or a theory of justice, i. e., a theory that would offer a basis for the critique of positive law, depending on whether or not it implements certain values. We are also not directly interested in accounting for judgments of recognition of an authority, such as the one used in premise *1)* of the argument which, a few pages above, we used in the discussion of Nino's thesis.[12] Rather, what we are interested in is a theory of legal sentences; thus, our approach to values is, so to speak, internal to legal sentences: the value judgments we are interested in are those that, more or less explicitly, constitute not the directive, but the evaluative dimension we believe legal norms do entail. For this, we think that we do not need to start from a general theory of values, although we must, first of all, reach a minimum of conceptual precision.

It is probably true that anything, any entity, can be subject to evaluation (Rescher 1969, 56 f.); but here we are interested exclusively in two objects of evaluation: actions and states of affairs. The reason for this is that they are the ingredients we need to account for normative sentences (besides, of course, deontic operators). Some states of affairs can be seen as (conceptually or causally) connected with actions. In the first case — when the connexion is conceptual or intrinsic — we will say that the state of affairs is the *result* of the action. In the second case — when the connexion is causal — we will call the state of affairs a *consequence* of the action (or actions) (cf. von Wright 1963, 39). For example, if X and Y satisfy certain characteristics (e. g., if they have a

---

[12] What we wish to underscore here is only that such judgments of recognition of an authority imply regarding the prescriptions coming from someone considered an authority not as pure prescriptions (whose incorporation as auxiliary reasons in the practical reasoning of their addressees depends, as in the case of the robber, on the operative reason constituted by some value judgment other than that of the prescriptor — like the value of one's own life), but as norms, that is, as operative reasons for acting that are guides for action as well as criteria for the evaluation of actions.

certain age) and perform certain activities which, according to the legal order, constitute the act of getting married, then this produces the normative result of making them married persons; the consequences can be that the two gain in emotional stability, that X's relationship with Y's mother improves, etc.

Value judgments attribute positive, negative, or indifferent value to the evaluated objects. To attribute positive value to an object (an action or state of affairs) means to assert that there is a reason for considering it justified to perform action A or to bring about state of affairs S. To attribute negative value to an object means that there is a reason for considering it unjustified to perform A or to bring about S. Finally, A and S are indifferent if there is no reason for considering the performance of A or the bringing about of S justified or unjustified.

The values incorporated in legal sentences can be regarded as the expression of value judgments, made by those who utter the sentences (the legal authorities), on certain actions and states of affairs. This does not mean that we attribute to the authorities (for instance, the legislator) the capacity to create values; rather, it means that we are interested only in the values adopted by legal authorities. Just as there are implicit norms, one can, of course, also speak of implicit value judgments. Those who accept such value judgments of the authorities also accept that there is a reason for thinking that it is justified (or unjustified, or indifferent) to perform certain actions or to bring about certain states of affairs.

From the perspective of the legal system, an action or state of affairs can be intrinsically or extrinsically valuable (or disvaluable). It is intrinsically valuable if it is attributed (positive) value for itself. It is extrinsically valid if the action or state of affairs is, by itself, indifferent (or even disvaluable) and if what makes it valuable is *exclusively* the connexion it is supposed to have with some action or state of affairs that is intrinsically valuable. Among the things that are intrinsically valuable, it is convenient to distinguish two subtypes: ultimate values and utilitarian values. To characterize some action or state of affairs as an ultimate value means that its consequences are disregarded; because if what made it valuable were its consequences, what is valued, in the last instance, would not be the action or state of affairs in question, but its consequences — i. e., other states of affairs. Thus, since here the relationship between actions and states of affairs can only be intrinsical or conceptual, that distinction looses much of its sense. If, for example, we look at freedom of expression — understood as a negative freedom — as one of those ultimate values, it makes no difference whether we say that what is valuable — or justified — are the actions that respect or guarantee freedom of expression, or the state of affairs in which freedom of expression is respected or guaranteed. This does not mean, of course, that freedom of expression has no consequences (for example, it makes the functioning of democracy possible, it facilitates social change, etc.); it only means that we value freedom of expression, no matter what are its consequences.

The difference between actions and states of affairs is, however, relevant with respect to utilitarian — i. e., intrinsic, but not ultimate — values. The characteristic of utilitarian values is that the actions and states of affairs regarded as such allow for a superior criterion of evaluation. Thus, for example, we positively value the state of affairs consisting in all citizens having either a job or some other source of income; but we value this basically because of its consequences: because it enables everyone to make and carry out a plan of life, etc. That is why, in general, we accept that this value is restricted by others, like that of equality: A policy of full employment that entailed some form of discrimination on grounds of sex would be considered unjustifiable, no matter how efficient — with respect to the value of 'full employment' — it would be.

All this does not mean that the sphere of ultimate values should be seen as coinciding with that of the values of freedom, whereas the sphere of utilitarian values should be identified with that of economic or social objectives. In order to reject this identification, we only need to observe that, according to the Spanish Constitution, freedom of expression, understood as a negative freedom, is an ultimate value, but understood as a material, or real, freedom, it is an objective restricted by the necessity to take into account other objectives and interests as well as, of course, the values constituting ultimate ends. Thus, for example, as we saw when we analyzed constitutional permissive sentences, the Spanish Constitution stipulates as one of the objectives to be pursued that individuals and groups should have the effective possibility of expressing their ideas and opinions. Also, with respect to health, it says that it is an ultimate value that a sick person should not be treated worse than other people who are otherwise in substantially the same conditions, whereas to offer the population the best possible health system is an objective that is not given the attribute of an ultimate value. In our view, this means that if we take each one of the major values to be found in our legal orders as a whole (health, liberty, equality, property, etc.), then their classification in terms of ultimate values and utilitarian values may not be possible. Rather, what happens is that concerning those major values, there are aspects which are seen as ultimate, and others merely regarded as objectives to be pursued (but which, although they are intrinsically valuable, are not ultimate ends).

Again, this does not mean that ultimate values have no bounds. Since legal orders do not recognize a single ultimate value and since it cannot be excluded that in some particular case, different ultimate values pointing into different directions are involved, we can say that those values restrict each other horizontally, but not vertically. Since they can clash with each other, the range of application of each one of them is restricted, but not their force or strength, because in that case they would not be ultimate values. Utilitarian values, in contrast, are not only restricted horizontally by other utilitarian values, but also vertically — with respect to their force or strength — by ultimate values.

As we have said, extrinsic values are of a purely instrumental kind (what makes such actions or states of affairs valuable is *exclusively* their relation with some intrinsically valid action or state of affairs). We can find that purely instrumental character in states of affairs considered valuable by the legal order as well as in states of affairs considered valuable by an agent himself. Thus, for example, to drive on the right is not, in itself, a valuable action; what makes it valuable — justified — is to see it as a mechanism (among others) for obtaining traffic safety (which, according to the legal order, is a valuable state of affairs). Punishments are, in themselves, disvaluable; what makes their imposition valuable is exclusively their character as a means of retribution (according to a retributive conception of punishment, it is an instrument for realizing an ultimate value) or general and/or special prevention (according to a preventive conception, punishment is an instrument for realizing a utilitarian value). The production of a certain normative result can be seen as instrumentally valuable with respect to certain ends pursued by those who use the power-conferring rule. Thus, for example, the making of a will is an instrument to be used by those who wish that their property, after their death, is disposed of in a certain way.

This already enables us to examine the different kinds of legal sentences we have distinguished, with respect to the values they entail or, in other words, from a justificatory perspective.

The justificatory aspect of principles is different in principles in the strict sence and in policies. The former provide an ultimate justification (since what they express are ultimate values), but not a conclusive one (until it is decided how possible conflicts with other norms are to be solved).

In the case of policies, the justification they provide is not ultimate, since they contain only utilitarian values, restricted — we can say, negatively — by ultimate values. It is also not conclusive, because a causally appropriate action for bringing about the state of affairs considered valuable from a utilitarian perspective by the policy can negatively affect the realization of other states of affairs also considered valuable from a utilitarian point of view by other policies. In contrast to principles in the strict sense, in policies the relationship between states of affairs and actions is extrinsic or causal. And that causal process is always, inevitably, affected by other causal processes which connect other ends (states of affairs) declared objectives to be pursued by the legal order to the actions that promote them. A justified action, we can say, is the action which, while respecting the other norms of the order (and especially the limits drawn by principles in the strict sense), is the most efficient, i. e., implies the smallest sacrifice of other ends. However, in the case of principles in the strict sense, we can determine a justified action irrespective of the causal process, that is, without considering its consequences. In other words, what we have here are criteria of correctness, not of efficiency. Critera of correctness imply an either-or requirement: a decision is or is not correct,[13] whereas criteria of efficiency can be satisfied in different degrees. For example, a decision may be more efficient than another one, but it makes no sense to say that a decision is more correct than another one (since in that case, the latter one would simply not be correct).

In the case of power-conferring rules, we must distinguish various elements. Action Y of the antecedent obviously entails a merely instrumental value: the action is valuable in so far as result R is valuable. State of affairs R can be evaluated from the point of view of the legal order, or from the point of view of the agents. From the former point of view, we can say that the legal order rules results to be obligatory because it considers them valuable (they may be valuable in an ultimate, a utilitarian or simply an instrumental sense), whereas what is facultative is considered to be indifferent. From the point of view of the agents, state of affairs R can be indifferent or can be considered valuable in an ultimate, a utilitarian or an instrumental sense. Also, one can make an instrumentally valuable use of power-conferring rules in two different senses: on the one hand, in so far as the subject is able use them adequately, in the sense of performing the actions appropriate for bringing about the result;

---

[13] This does not exclude — but neither does it imply — the possibility of *tragic cases*, that is, cases for which there is no correct answer within the system.

and on the other, those norms can also be used as instruments for attaining social and individual consequences, that is, as instruments serving values or ends that are external to the legal order.

Mandatory rules are specifications of the generic circumstances constituting the conditions of application of values of any one of the kinds we have distinguished. This means that the values incorporated in the actions or states of affairs deontically qualified by rules as obligatory or prohibited can be of the ultimate, of the utilitarian or simply of the instrumental kind (or consist in some combination thereof). Thus, the norm prohibiting to kill another person (unless there are certain special circumstances) contains an ultimate value. The norm commanding that in urban areas a certain percentage of the territory must be reserved for parks or recreational facilities contains a utilitarian value. And the norm obliging one to pay a certain percentage of one's annual income, depending on the amount of that income, as income tax obviously contains a utilitarian value, but also ultimate values in the form of constraints on the realization of that utilitarian value (since, e. g., fiscal pressure must be progressive relative to income, it must respect the prohibitions of discrimination contained in art. 14 of the Spanish Constitution, etc.). And the norm ordering one to drive on the right, or a norm ordering an administrative organ to complete certain tasks within certain time limits, simply contain an instrumental value.

Concerning permissive sentences, remember the distinction drawn in ch. III, depending on whether such sentences appear in the context of principles, of power-conferring rules, or of rules regulating 'natural' conduct. With respect to the first two of these contexts, we have already indicated what their value aspect consists in. As for rules regulating 'natural' conduct, it will be recalled that, from the directive point of view, we characterized permissions in this context either in negative terms (as anullment of, exception from, or clarification of the domain of, prohibitions) or in terms of an indirect formulation of prohibitions. From the second perspective, we have nothing to add to what has been said about mandatory rules. And from the first, we can say that the evaluative dimension of permissions in this context must also be characterized in a negative way: In this context, permissions imply that the actions in question are not (or no longer) subject to evaluation by the legal order (and, therefore, their evaluation is left to the subject's own value judgments).

CHAPTER V
THE RULE OF RECOGNITION

1. Introduction

So far, we have been occupied with different kinds of legal sentences. The purpose was to clarify what kinds of pieces the machinery of the law consists of. In this chapter, our aim is rather to examine that machinery as a whole. As is commonly known, in the legal theory of the positivist tradition, the unity of the legal system has been seen as a correlate of the possibility to refer each and every one of its provisions to a single 'master norm' — to use the expression coined by Dworkin (1978) — located above the norms issued by the system's authorities. That 'master norm' is understood in different ways, and it has also been given different names, of which those of 'basic norm' *(Grundnorm)* (Kelsen) and 'rule of recognition' (Hart) are the most famous. As is also well known, for Kelsen, this norm which is presupposed by legal theorists, states "In short: One ought to behave as the constitution prescribes" and is "the condition under which the subjective meaning of the constitution-creating act, and the subjective meaning of the acts performed in accordance with the constitution, are interpreted as their objective meaning, as valid norms " (Kelsen 1967, 201 and 204). In Hart's version, in turn, it is the norm providing the ultimate criteria of legal validity and which exists "as a complex, but normally concordant, practice of the courts, officials, and private persons in identifying the law by reference to certain criteria" (Hart 1994, 110).[1]

---

[1] The common problem underlying Kelsen's notion of the 'basic norm' as well as Hart's notion of the 'rule of recognition' is that of the unity of the legal system and of the obligatoriness of its ultimate sources. Kelsen's notion, however, points to a norm *presupposed by legal theorists* and which is a *condition of the possibility of a pure legal science*, i. e., which excludes "everything that is not strictly law" (Kelsen 1967, 1) — e. g., ethics, politics, and value judgments in general, on the one hand, and causal sentences, on the other. Hart's notion, in turn, points to a norm that *exists insofar as it is socially accepted* and that is the *condition of the existence* of a legal system. Still, one should not forget that, according to Kelsen, it only makes sense to presuppose the basic norm with respect to systems of norms that are, as a whole, socially effective. Thus, both constructions are answers to — or, if you prefer, conceptions of — the central problem common to all the different versions of conceptual legal positivism: that of the *social* roots of that *normative* system we call 'the law'. As will be seen, the following exposition is more in line with Hart's answer. On some of the deficiencies of Kelsen's principle of effectiveness, cf. Ruiz Manero 1991.

2. Jurists and the 'normative value' of the constitution

In contrast to what one may think, that notion of a 'master norm' is not simply an 'invention' of legal theorists. Of course, academic lawyers usually do not invoke norms located above those contained in the source they identify as supreme (norms, that is, which would, in a certain sense, be more basic than these). But the fact that the question cannot be avoided — and that legal theorists do not insist on it merely to entertain themselves, because they enjoy talking nonsense or multiplying entities unnecessarily — becomes apparent when academic lawyers start to speak of the obligatoriness of the norms contained in the supreme source. This can be seen very clearly when one looks at how some of our most renowned jurists have approached the topic of the 'normative value' of the Spanish Constitution of 1978. Let us take as a starting point the work of one of our best constitutional theorists. In his book *Derecho constitucional. Sistema de fuentes (Constitutional Law. The System of Sources)*, Ignacio de Otto writes the following:

"Of course, the issuing of a supreme norm, above the highest organs of the State, is done by enacting a written text, the so-called written constitution, whether it is given the name of Constitution or some other name; but there is a constitution as a norm only when the order stipulates that it is obligatory to comply with those prescriptions and that, therefore, their violation is unlawful." (De Otto 1989, 15)

"If the order does not contain such a provision, i. e., if the violation of the written constitution is not unlawful, then the provisions of that constitution are constitutional only in the sense that they are included in the constitution, but in fact they are not even norms, because a norm that may be violated lawfully *is no norm*. On the contrary, if the order stipulates that it is obligatory to obey the written constitution, *all* its provisions are equally obligatory, whatever matter they may treat, and all of them have the characteristic of a supreme norm. What makes a norm a constitutional norm is that the order attributes a supreme position to it, placing it hierarchically above legislation." (Ibid., 18)

Obviously, the author cannot explain the conditions that confer truth to the assertion that the Constitution of 1978 is the supreme source of the current Spanish legal order, and he takes refuge in the formula that it is "what the order stipulates". Since, apparently, reference to the legal order can point to nothing but the norms that order contains, his assertion can be interpreted in two ways. The first is that when he speaks of the 'order' he refers to some norm belonging to the constitution. The second possibility is that it is a norm apart from the constitution itself. In a work whose influence on Spanish legal culture can hardly be exaggerated, Eduardo García de Enterría has used both of these ways for grounding the assertion that compliance with the constitution is obligatory. Thus, he writes:

"[...] our Constitution explicitly attributes a normative, directly binding value to the constitution" (García Enterría 1988, 61)

"The first thing one must state with absolute clarity is that the entire Constitution has an immediate, direct normative value, as can be deduced from article 9.1: 'The citizens and the public powers are subject to the Constitution and the rest of the legal order." (Ibid., 63)

"This interpretation is corroborated, as in a true borderline test, by the most conspicuous case of the strictest subordination to hierarchical organization and orders: that of persons belonging to the military organization; the Royal Ordinances for the Armed Forces, enacted on December 28, 1978, in their article 34 explicitly stipulate that the obligation of compliance with the orders of a superior for all members of the Armed Forces [...] has the following limit: 'when the orders entail the execution of acts manifestly contrary to the laws and customs of war, or if they are a crime, especially against the Constitution'." (Ibid., 64)

Thus, implicitly in Ignacio de Otto and explicitly in García de Enterría, we find two lines of reasoning for a foundation of the obligatoriness of the Constitution: the first is that the Constitution is obligatory because the Constitution itself says so; the second is that the Constitution is obligatory because a norm belonging to the legal order, but not to the Constitution itself, says so. Obviously, neither one of the two ways can found anything at all. As for the first, it clearly commits a *petitio principii* since it presupposes that the Constitution is obligatory. And as for the second (that the Constitution is obligatory because this is what some other norm, apart from the Constitution) says, it is obvious that it depends on a norm which is said to be obligatory and belonging to the legal order because it has been issued by the organ and through the procedure established in the Constitution, and therefore, it cannot be used to found the obligatoriness or the belonging to the legal order of the Constitution itself.

What neither Ignacio de Otto nor Eduardo García de Enterría seem to have noticed is that the question of how to identify the supreme norms of a legal system is a problem that cannot be solved from within the system itself: any attempt in that direction either leads to a circular argument (the Constitution is obligatory because that is what the Constitution itself says, or what a norm says which, itself, is obligatory only according to the Constitution) or to nothing (the obligatoriness of the Constitution is asserted without offering any reason at all; the Constitution is said to be obligatory because it is).

Besides, it is strange that García de Enterría tries to present as a *fact* what, at the time of his writing (the first version of his work is from 1980), was only a *proposal*. What García de Enterría proposed in 1980 is *that the normative claims of the Constitution be recognized*.[2] Thus, in the Preface, where he explains the genesis and motivations of his work, he writes:

"The task was to make possible, with great urgency, the application of a constitution born with an explicit normative intention but, nevertheless, threatened by our tradition which, from the beginnings of the constitutional period, was never suspended and which regarded constitutional provisions as mere 'program norms', addressed to the legislator simply as non-binding recommendations, and with no effect of their own on citizens and judges. I like to think that my investigation could contribute in some way to correct that impressive tradition." (García de Enterría 1988, 33)

Of course, Enterría's work has been successful. But that is so because our judiciary[3] and the Spanish legal community as a whole have accepted a norm — necessarily located outside of the legal system (in the sense that it is not prescribed by any authority of the system) — that commands the recognition of the constitution as the supreme source of that system, and have agreed not to understand it, as used to be the case in the predominant tradition of Spanish constitutional jurisprudence, merely as a political document. And it is this fact of a shared acceptance of an unwritten norm, and not any constitutional statement, that today gives truth to the proposition that 'The Constitution of 1978 is the supreme source of the Spanish legal system'. As late as 1982 (decision of April 8), our Supreme Court has denied direct normative value to the Constitution, regarding it merely as a document to guide the public powers, and espe-

---

[2] Our critique here is restricted to García de Enterría's treatment of the problem of the recognition of such normative claims. Once they have been recognized, obviously, the constitution must be understood to be directly applicable by judicial organs (without any prior *interpositio legislatoris*) and to stipulate the subjection of political powers to the law and, especially, to the constitution itself. While these two points have been brilliantly solved by García de Enterría and have as such become part of our common constitutional wisdom, this is not the case with the first problem which is the one we are interested in here.

[3] Here and elsewhere in the text, we use the expression 'judiciary' and other, similar expressions (like 'judges and courts', etc.) in a loose sense, encompassing not only the judicial organs belonging to the 'third power', but all organs "concerned with the authoritative determination of normative situations in accordance with pre-existing norms" — to say it in the words of Joseph Raz (1990, 134) —, as for example (to mention two very different examples) the Constitutional Court and the so-called economic-administrative courts (which are organs not of the judiciary, but of public administration).

cially the legislative power whose intervention was regarded as absolutely necessary to transform constitutional standards into legal norms, that is, norms able to serve as a foundation for the decisions of the courts. Suppose that line of jurisprudential reasoning had quietly taken hold of all our law-applying organs. In that case, the proposition that 'The Spanish Constitution of 1978 is the supreme source of the Spanish legal system', understood as a proposition about the legal system actually in force, would clearly be false. Despite of its theoretical shortcomings, García de Enterría's work, without any doubt, has decisively contributed to make that proposition true today.

3. The rule of recognition as ultimate norm

From all this it follows that the identification of the supreme source of a legal system depends on a norm existing only insofar as it is commonly accepted. In the Hartian tradition of legal theory, that norm is called the *rule of recognition*. Precisely because its existence depends merely on its shared acceptance, such an ultimate norm, such a rule of recognition, is not a legal sentence like all the others; and if we limit legal sentences to sentences produced in accordance with the system, it is not even a legal sentence at all: the rule of recognition can only be expressed in a meta-language, and cannot be part of the object-language, that is, the language of positive law. Thus, one cannot say that the rule of recognition is a norm — or set of norms — of the order (the constitution) that is given greater weight than the others. Apparently, that is what Gegorio Peces-Barba (1984) had in mind when he identified (albeit with some 'nuances' which, however, are not sufficiently elucidated) article 1.1 of the Spanish Constitution ('Spain is constituted as a social and democratic State under the rule of law, declaring as the highest values of its legal order liberty, justice, equality and political pluralism') as the rule of recognition of Spanish law. In his own words:

"As a material basic norm, [art. 1.1] is the norm of identification of the other norms, with respect to their material content. That is, a norm is considered to be of the order, is identified as part of the order, if it endorses — positive perspective — or does not contradict — negative perspective — the highest values. What the material constitution adds to the identification of norms in legal orders is precisely that that identification is not only produced by formal criteria — a competent organ and an adequate procedure for producing the norm —, but also by criteria of content.

Article 1.1 is the basic norm of material identification of the order (what Hart calls 'rule of recognition', with some nuances, giving it a material sense)." (Peces-Barba 1984, 97)[4]

Now, in our view, what this author has confused here is the concept of ultimate norm, or rule of recognition, and the concept of materially most important norm. Article 1.1 may be the materially most important norm of the Spanish legal order,[5] but it does not, of course, stipulate the *ultimate* criteria of legal validity, because its own identification as a legal norm depends precisely on the previous recognition of the Constitution as the supreme source of the system.

4. *Changing the rule of recognition without rupturing legal continuity?*

To what has just been said, the following objection can be raised: It is wrong to assume that the Constitution is directly pointed out as the supreme legal source by the accepted rule of recognition; the Spanish Constitution of 1978 was enacted by a parliament — the *Cortes* — elected in accordance with the Law for Politial Reform which, in turn, was issued in accordance with the Basic Laws of the Franco regime, and so on, so that legal continuity would go back all the way to the provisions of the Junta of Burgos which appointed General Franco Chief of State. Thus, the accepted rule of recognition would lead us to the 1936 Junta of Burgos, rather than to the Constituent Assembly of 1978, as the supreme authority.

Although it may reflect mental habits and patterns of interpretation not uncommon among jurists, in our opinion this objection is clearly mistaken. Be-

---

[4] Peces-Barba's complete thesis is that the basic norm consists not only of the highest values, but also of the sovereignty from paragraph 2 of that same article 1: 'National sovereignty belongs to the Spanish people from whom the powers of the state are derived.' The two parts of the article are said to be complementary in the following way:
"If we take a closer look at the answer to the questions 'Who commands?' and 'How does he command?', that is, at sovereignty and the rule of law, we are facing the kind of legal organization Kelsen had in mind when he constructed his theory of dynamic orders, that is, the basic norms of the formal constitution, whereas if we look at the answer to the question 'What is commanded?', that is, at the highest values, we are facing the basic norm of the material constitution. They are inseparable, and both perspectives make up the basic norm, articles 1.1 and 1.2 of the Spanish Constitution" (Peces-Barba 1984, 93).
"The formal basic norm — popular sovereignty and the rule of law — and the material basic norm — liberty, justice, equality, and political pluralism — express one and the same ideological conception which is the foundation of the constitutional consensus of the Spanish people. The first reflects the legal structure of a representative-parliamentary state, the second the theory of justice, the material contents of a representative-parliamentary state order" (ibid., 94).
[5] This may be controvertible, but we do not want to discuss it here. In any case, cf. Laporta (1984) and Ruiz Miguel (1988).

cause, though every irregular modification of the supreme source — if it is successful — translates into the use of a new rule of recognition by the courts and officials, not every regular change of the supreme source implies continuity of the same rule of recognition. An already classical example is the one advanced by Hart (1994, 120 f.), of the colonies which acquired independence through a constitutional structure established by an act of the Westminster Parliament. Once independence is consolidated, the rule of recognition accepted in a former colony no longer includes any reference to the legislative powers of the British parliament, even though the local constituent authority was originally created by an act of that parliament. Another example in the same direction is that of Kent Greenawalt (1988), concerning the relation between the U. S. Constitution and its clause of ratification. That clause stipulates that the Constitution will come into force in those States that have ratified it when nine of them have done so. Thus, Greenawalt says, one could think that the rule of recognition of the U. S. refers directly to that ratification clause, and not to the Constitution as such, which would count as valid law by derivation from the validity of the ratification clause. This example, obviously, is somewhat more complicated than the one before, because the ratification clause did not have legal status prior to the rest of the Constitution that was to be ratified according to it. But, in any case, the decisive argument for sustaining that the rule of recognition accepted today in the United States refers directly to the Constitution as a whole, and not simply to its ratification clause, is that today nobody would consider it a legal argument against the validity of the Constitution to say that some State did not ratify it in the right way. In our view, the same applies to the Spanish case: If someone would argue against the validity of the Constitution of 1978 saying that, for instance, the referendum about the Law for Political Reform of 1976 had been rigged, this would be regarded as merely of historical interest; as a legal argument, it would not be taken seriously.

5. A host of problems

The rule of recognition, thus, is a norm existing only — as Hart says in a passage already quoted earlier — "as a complex, but normally concordant, practice of the courts, officials, and private persons in identifying the law by reference to certain criteria" (Hart 1994, 110). Those criteria are the *ultimate* criteria of legal validity; thus, the rule of recognition "provides criteria for the assessment of the validity of other rules; but [...] there is no rule providing criteria for the assessment of its own legal validity" (ibid, 107). Therefore, the rule of recognition is neither legally valid nor invalid, because there is no other, higher criterion of legal validity. This, of course, gives rise to a number of questions which can be grouped into the following blocks:

a) It has just been said that the rule of recognition exists "as a complex practice". But who are the subjects of that practice? And what are its characteristics? Are the different kinds of subjects, like judges, legislators, and private persons involved in different practices, all of them contributing to shaping, maintaining, or gradually changing the rule of recognition? Do some of these practices, or some of those types of subjects, in some sense have a privileged position?

b) What does the rule of recognition prescribe that is not prescribed in the norms identified according to it, that is, the norms of the legal system? Is the rule of recognition really a norm, or is it only a conceptual criterion for the identification of norms? What functions does the rule of recognition fulfil? Does acceptance of the rule of recognition necessarily require a moral justification?

c) Does every legal system have one and only one rule of recognition? Can the ultimate criteria of legal validity contained in the rule of recognition contain indeterminacies or zones of penumbra?

We will try to answer each one of these blocks of questions, one at a time, in the following paragraphs.

6.   Who shapes the rule of recognition?

Concerning the first block of questions, in our view, we must start from the following idea: When it comes to identifying the rule of recognition of some legal system, from an external point of view, we must, in the last instance, look at the criteria of legal validity accepted in the practice of the courts; but, of course, that does not mean that the judiciary is the authority that issues the most basic norm of the system. What we wish to underscore is the decisive importance of the acceptance of that norm by judges and courts — an acceptance which implies that they also recognize their duty to apply pre-existing norms in their rulings (that is, that they recognize the authoritative character of such norms, or if you prefer — in the case of explicitly prescribed norms which play a central role in all developed legal systems —, the authority of those who have enacted them). Let us now take a closer look at all this.

6.1. The institutional role of a 'judge' or 'court' seems to have two defining elements: the first one is having the normative power to solve the cases in an authoritative way, that is, the normative power to issue binding decisions on the cases brought before them; the second one is having the duty to see to it that such decisions are applications of pre-existing norms which are binding for the judge or court himself. If one of those two requirements is missing, we are not dealing with a judge or court in the legal sense (that is, the sense of a law-applying organ), but with a referee, an umpire, a mediator, or something of the kind. The sum of those two characteristics explains why it is crucial to look to the courts when it comes to determining the legal system existing in a community (or, in other words, the rule of recognition allowing one to identify the set of norms that legal system is composed of). Because it is, in fact, the courts which are authorized to take binding decisions on the normative situation of individuals. Thus, if the norm-creating institutions would come into conflict with the norm-applying institutions (with the courts) and the latter would not recognize as binding the norms issued by the former, then what would be relevant for the general public would be considerations about those norms actually recognized by the courts as binding, and not about those the (alleged) norm-creating authority enacts with the mere claim — not recognized by the courts — of bindingness. So, we could say that an (alleged) norm-creating authority, an (alleged) legislative chamber or constituent assembly, is such a thing if and only if it is recognized as such by the courts. Going back to the Spanish case, to say that the normative claims of the Constitution have become reality, that the Constitution is, in fact, the supreme source of the system, basically (if understood as an assertion about existing law) means that the courts have recognized it as such.

6.2. This recognition of the Constitution as the supreme source of the system by the courts can be seen as the result of a number of different factors, among which a general attitude of acceptance towards the Constitution's claim to normative supremacy among jurists, among the political class, and among the general public seems to be fundamental. That general attitude, in turn, can be regarded as resulting from the belief in the value of the Constitution for giving the country a framework for a stable life in society respecting certain individual rights said to be of basic value, etc. All this takes us to two kinds of questions to be answered: first, that of the reasons underlying the acceptance of a rule of recognition; second, that of the contribution of different kinds of subjects, besides the judiciary, to that situation of shared acceptance the existence of a rule of recognition depends on. We will leave the first question for later treatment, and take on the second one — that of the contribution of subjects like legislators and administrative organs, lawyers and the proverbial 'man in the street' to the existence of the rule of recognition.

6.2.1. The contribution of legislators and norm-producing administrative organs to the existence of the rule of recognition basically translates into their acceptance of the normative framework that gives them their powers to produce norms and imposes duties concerning the exercise of those powers on them (for example, imposing constraints for the possible content of legislation). Such acceptance implies the use of the conferred powers, and compliance with the related duties (for example, the respect of constitutional constraints on the content of laws).

6.2.2. Concerning legal theorists, the example of García de Enterría mentioned earlier is instructive about the capacity they may come to have for contributing to the shaping of the rule of recognition accepted by the judiciary. As for practicing lawyers, their contribution to that existence can be seen to consist mainly in forming expectations: if they would not 'expect' judges, legislators, and administrative organs to act — at least up to a certain point — in accordance with the system's rule of recognition, their own practice as lawyers would make no sense.

6.2.3. Finally, even mere private citizens contribute to the existence of the rule of recognition, because one could not use the law as a mechanism of foreseeing the consequences of one's own behaviour, if one would not expect norm-applying and norm-producing organs to follow one and the same rule of recognition. Therefore, when the latter seem to deviate from it, public opinion — insofar as it does not desire a change of the rule of recognition — reacts with criticism. An interesting example of this seems to be the case of certain critiques, published in the media, of acquittals in some cases of 'conscientious objection' (against military service). We are referring to critiques reproaching the judge for having put his own conscience above legislated norms which are binding for him, that is, to have acted against what is required by the rule of recognition. We will come back to this case, which will also serve to illustrate another problem related to the rule of recognition.

6.3. We have said that the ultimate criteria of legal validity allowing us to identify the set of norms a legal system consists of are part of a norm — the rule of recognition — that exists only insofar as it is accepted and practiced by the entire judiciary (although, as we have seen, other subjects also play a role in shaping and maintaining that ultimate norm). This seems to pose the following problem: In order to determine what the rule of recognition of a legal system is, we must first identify what organs the judiciary of that system consists of; but those organs, in turn, are what they are because of the rules of that system which confer judicial powers, and the validity of such rules depends, in the last instance, on their being in accordance with the rule of recognition of the system. Therefore, putting a rule of recognition accepted by the judiciary at the base of a legal system seems to lead into a circle: in order to determine the rule of recognition, we must identify the judiciary; but we cannot do that without taking into account rules conferring judicial powers whose validity, in turn, in the last instance, depends on the rule of recognition. Several solutions intended to 'break the circle' by identifying the judiciary with the help of criteria other than the rules of the system conferring judicial powers have been suggested for this problem. Thus, Neil MacCormick has proposed to identify judges in terms of social rules of duty (MacCormick 1981, 109 ff.); Carlos Nino, in contrast, has proposed to go about it in purely factual terms, characterizing as judges and courts "those who *actually can* (in the factual, not in the normative sense of the word 'can') determine the exercise of the state's monopoly of coercion in particular cases, that is, those who are, in effect, able to set in motion the coercive apparatus of the state" (Nino 1980, 128). Elsewhere, one of us has already criticized these proposals (cf. Ruiz Manero 1990, 124 ff.), and we do not need to repeat that criticism here. But what we do wish to point out is that these attempts to 'break the circle' which, besides, were highly artificial — just like some of the detours in Juan Ruiz Manero's attempt (ibid.) to present an alternative solution —, overestimate the difficulty of the problem. Because we can do with a much simpler criterion for identifying judges and courts: we can just look at those who are socially recognized as such, that is, as holders of the powers and duties defining, as we have seen, the institutional position of the judiciary. And for identifying judges in this way apparently we do not need much theoretical sophistication or special knowledge; it is purely a matter of everyday experience. If one accepts this point of departure as the central pillar of our 'model' of the legal system, the fact that this 'model', seen from within, becomes 'circular' — in the sense that in order to determine the accepted ultimate criteria of legal validity (that is, the rule of recognition) we refer to the practice of the judiciary, and in order to determine who belongs to that we refer to rules whose validity depends, in the last instance, on those accepted ultimate criteria — does not pose any special problem; because, as Ricardo Guibourg correctly remarked, "seen from within, any system [is] circular" (Guibourg 1993, 431).

7. *The conceptual, directive and evaluative dimensions of the rule of recognition. The rule of recognition and the exclusionary claim of the law. Why accept the rule of recognition?*

We said that the central pillar of the rule of recognition of the Spanish legal system is the acceptance of the Constitution of 1978 as that system's supreme source. Thus the rule of recognition refers directly to the Constitution, and indirectly to the norms issued or received in accordance with it. Or, to express the same idea in terms of authorities — at the cost of not being able to account for norms that do not have their origin in acts of an authority, as, for example, customary norms —, the rule of recognition refers directly to the authority of the constituent assembly, and indirectly to the authorities recognized or instituted by it. All this means that the rule of recognition, on the one hand, provides a criterion for the identification of norms and, on the other, is a guide for behaviour and a criterion of evaluation, for the general public as well as for the law-creating and law-applying organs.[6] As a criterion for the identification of norms, the rule of recognition draws the limits of the Spanish legal order: the norms belonging to that order are those contained in the Constitution and in the sources recognized or instituted by the Constitution. As a guide for behaviour and a criterion of evaluation, it commands that the norms thus identified should be obeyed, and it also refers to those norms as a criterion for the evaluation (i. e., the justification or critique) of behaviour.[7]

---

[6] In common usage, 'recognition' can be linked to practical contexts (of behavioural guidance: 'I recognize the authority of so-and-so', or of evaluation: 'I express to you my recognition for the work you have done') as well as to theoretical contexts ('I recognize the truth of that proposition', 'I recognize this metal to be copper'). From this perspective, we think it has been especially appropriate to call rule of *recognition* a norm which, besides the practical dimension it has as a norm (as a guide for behaviour and a criterion of evaluation) also offers an ultimate theoretical criterion for the identification of legal norms.

[7] By referring to a certain ultimate source, the rule of recognition enables us to identify the *independent norms* of a legal order. In legal theory, the expression 'independent norms' has been coined for those norms whose validity, or belonging to a legal order, is predicated directly from the rule of recognition, and not from criteria of validity or belonging provided by other norms (cf. Caracciolo 1988). That means that the norms contained in the ultimate source the rule of recognition of a legal order refers to are independent norms of that order. Based on this, the rule of

Depending on the context, the two practical dimensions — that of a guide of behaviour and and that of a criterion of evaluation — emphasize different things. Thus, of the norms the rule of recognition refers to, those addressed to the general public — which, following Alchourrón and Bulygin (1971), we can call the *primary* or *subject system* — function mainly as guides of behaviour for their addressees (the general public), whereas for the law-applying organs they are criteria of evaluation that must be used to judge the behaviour of the former. As for common private citizens, they too use the rule of recognition and the norms it refers to as criteria of evaluation, either of the behaviour of other private citizens or of the legal decisions issued by the norm-applying (or norm-creating) organs. And as for the law-applying organs, the norms addressed to judges as such operate for them mainly as guides of normative behaviour while, at the same time, providing the criteria of evaluation according to which higher courts — formally — and the whole of the legal community — informally — judge such normative behaviour. The difference between organs of application and simple private citizens, in that respect, seems to be that the former, but not the latter, are addressees of norms, functioning for them as guides of behaviour, which order them precisely to use other norms of the system as criteria of evaluation for the solution of the cases presented before them.

7.1. What has just been said seems to give rise to a first difficulty which we can call the 'redundant character of the rule of recognition' and which affects the practical dimensions of that rule (that is, its dimensions of a guide of behaviour and a criterion of evaluation). Because if the rule of recognition, as a guide of behaviour and a criterion of evaluation, refers to the norms contained in the Constitution and in the sources recognized or instituted by it, then what guide of behaviour and criterion of evaluation does the rule of recognition provide that is not already contained in the norms it refers to? This circumstance, that

---

recognition of any legal order can be presented in one and the same *canonical form*. As a guide of behaviour, it would say 'Independent norms and norms issued or received in accordance with them ought to be obeyed'; and as a criterion for evaluation, it would say 'Behaviour and decisions ought to be evaluated (i. e., considered justified or unjustified) only on the basis of independent norms and the norms issued or received in accordance with them'. Finally, if what we want is the criterion for the identification of legal norms provided by the rule of recognition, then the formulation would be 'Legal norms are all independent norms and all norms issued or received in accordance with them'. If the legal system in question has more than one ultimate source — as is the case, for example, in English law (cf. below, 8.1) —, the formulation of the rule of recognition must also reflect the rank order of those different sources (and of the sets of independent norms contained in each one of them).

the rule of recognition does not order or justify anything not already ordered or justified by the norms it refers to, has led Eugenio Bulygin to think that, as a norm, the rule of recognition is totally superfluous, and that it should be understood merely as a conceptual criterion for the identification of norms, that is, as a definition (Bulygin 1976, 1991a, 1991b; cf. on this also Ruiz Manero 1990 and 1991).[8] Now, in our opinion, the obligation, stipulated by the rule of recognition, of following as guides of behaviour and using as criteria of evaluation the norms contained in the sources it refers to (i. e., the valid norms of the system) is not merely a repetition of the content of those norms. What the rule of recognition commands is that, whenever applicable, the conventions and commands of authorities the rule itself refers to — and not what the addressee himself, on the basis of his own balance of the reasons pertinent to the case in question, considers to be the best course of action to follow, or the best decision to issue, or the most fitting evaluation of the behaviour to be judged — ought to be *accepted as norms* (that is, followed as guides of behaviour, and used as criteria of evaluation). In ch. I, we have accepted Raz's (1990) thesis that the legal system is an *exclusionary system*, in the sense that, when its norms are applicable, it requires that one act and judge on the basis of those norms, excluding all reasons other than those norms as a basis for action and evaluation (except in those cases where the norms themselves permit the consideration of other reasons). Although we will later discuss some of the characteristics of this thesis in Raz's work, what we are interested in now is that to accept the rule of recognition is equivalent to accepting the law's *claim of being exclusionary*.[9] This makes it possible that (in contrast to a mere conceptual criterion) the rule of recognition can be disobeyed, although disobeying the rule of recognition also necessarily implies disobeying some other norm of

---

[8] Bulygin thinks that the rule of recognition should be interpreted merely as a conceptual rule (i. e., as the set of criteria for the identification of valid law, that is, as a definition of 'valid law'), and not as a social norm (which would command, generally, to obey the norms identified with the help of those criteria, i. e., legal norms and, particularly, that judges use those norms as the foundation of their decisions). Bulygin sustains that to consider those duties to arise from the rule of recognition would simply be redundant, since they are already stipulated by the norms of the system we identify in accordance with the rule of recognition: If a norm N_1 obliges one to do p, then nothing would be added by another norm N_2 which stipulates only that one ought to obey N_1.

[9] In the first chapter, we said that what distinguishes *rules* from *principles*, within the class of mandatory norms, is that the former are *exclusionary reasons*, while the latter are merely *first-order reasons*. Since the ultimate norm we have called rule of recognition is *an exclusionary reason in favor of the legal system as a whole*, we think it is particularly appropriate to call it a *rule*.

the system, just as one cannot obey the rule of recognition without obeying some other norm of the system.

Therefore, in our opinion, the thesis formulated, among others, by Moreso, Navarro and Redondo (1992) that for a legal justification one does not need to go beyond that norm of the system that works, so to speak, as the major premise of the well-known syllogism is untenable. Those authors believe that an argument like the following:

1) If z is a resident of district A, he ought to pay tax I.
2) z is a resident of district A.
3) z ought to pay tax I.

is a sufficient (complete) justificatory argument if it is shown that *1)* is a legal norm. And that, they say, can be done by the following theoretical argument:

1') There is a sovereign norm[10] authorizing districts to issue norms on municipal taxes.
2') District A has issued the norm contained in *1)* on municipal taxes.
3') The norm contained in *1)* is a legal norm.

Thus, what they are saying is that in order to justify a judicial decision, it *suffices* to understand the rule of recognition as a conceptual criterion for the identification of legal norms. In our opinion, however, that is wrong. In the previous argument, from the set of *1')*, *2')*, and *3')* one cannot simply jump to *1)*. The conclusion of the second (theoretical) argument 'The norm contained in *1)* is a legal norm' is *relevant* for thinking that z ought to pay tax I, regardless of any reasons that would support the opposite decision, or that z acts in a justified way if and only if he pays tax I, only if one presupposes another premise, namely: 'As guides of behaviour and critera of evaluation, one ought to follow legal norms and not any other, possibly conflicting reasons'. That is, one must presuppose the rule of recognition as a norm referring to legal norms as guides of behaviour and criteria of evaluation, excluding any other reasons that may be applicable to the case.

---

[10] The authors use the term 'sovereign norm' for referring to the concept which we have prefered to call an 'independent norm' (cf. above, n. 6); the expression 'sovereign norm' is taken from von Wright (1963, 199).

7.2. The thesis that legal systems claim to exclude any other, countervailing reasons — in the sense described above — and that acceptance of the rule of recognition implies acceptance of that claim of being exclusionary must be distinguished from the thesis — closely connected to it in Raz's work — that legal systems "do not acknowledge any limitation of the spheres of behaviour which they claim authority to regulate", since "they claim authority to regulate any type of behaviour" (Raz 1990, 150). In Raz's view, this is also true of those legal systems "which contain, for example, liberties granted by constitutional provisions which cannot be changed by any legal means. Such systems may not claim authority to regulate the permitted behaviour in any other way but they regulate it in one way by permitting it" (ibid., 151). In our view, that thesis is obviously confused, and the confusion is rooted, we think, in the notion of 'claiming authority to permit'. To claim authority (exclusively) to permit a certain form of behaviour without, at the same time, also claiming authority at least to prohibit it,[11] is not to claim any authority at all over that form of behaviour. Because one cannot claim authority over a form of behaviour without claiming the legitimacy of providing guidance on that form of behaviour. But permission, in contrast to commands or prohibitions, does not by itself offer any guidance. Thus, if the constituent assembly declares that it cannot legitimately regulate some form of behaviour except in the form of a permission, it is not asserting a claim of authority over that form of behaviour; rather, it is acknowledging a limit to that claim, that is, it points to an area over which it does *not* claim authority. That is the case when the Spanish constituent, for example, *acknowledges* — according to the formula repeatedly used in our Constitution — certain rights of freedom. Over these matters, the constituent does not claim authority, but asserts that there are domains where it cannot legitimately intervene with acts of authority.

7.3. Thus, even after this relativization, as we saw before, the existence of a rule of recognition of a legal system still consists in a shared acceptance of the norms that rule refers to as guides of behaviour and criteria of evaluation; and whenever they are applicable, those norms exclude the applicability of any other reasons not contained in, or recognized by, them. The rule of recognition thus exists as a conventional[12] (or, if you prefer, a customary)[13] norm that, in

---

[11] On this point, see Guibourg/Mendonça (1995).
[12] Juan Carlos Bayón defines a 'convention' in the following way: "A regularity R in the behaviour of the members of a group G in a recurring situation S is a *convention* if and only if in (almost) all cases of S

the case of developed legal systems, refers mainly to norms issued by authorities (in the Spanish case, directly to the Constitution, and indirectly to the norms issued or received in accordance with it). If understood in this way, the rule of recognition provides the ultimate criterion of legal justification, because beyond it, there are no legal criteria of any kind. And the only thing a legal system requires is a justification in terms of the system itself: For the law, it suffices that one follow the rule of recognition, irrespective of the reasons one may have for doing so.[14] However, from the point of view of the practical reasoning of an individual who follows the rule of recognition, the question about the reasons justifying this compliance makes perfect sense. And a person asked to justify her compliance with some rule of recognition, if she does not want to commit a circularity (and a naturalistic fallacy), cannot invoke as an operative reason for it the fact that there is such a convention, nor the fact that there exists a prescription from the authority the rule of recognition itself refers to as the supreme authority; instead, she must give *autonomous* reasons, i. e., reasons valid because of their intrinsic merits (and which, in any case, can give some value, as auxiliary reasons,[15] to conventions or acts of authority with certain characteristics). Such reasons, besides, cannot be based on that person's own interest, but must be *impartial*, because the norms the rule of recognition refers to (i. e., legal norms) obviously impose duties not only on the interrogated person herself, but also on all others who fulfil its conditions of application. But reasons that are, at the same time, autonomous and impartial we usually call 'moral' reasons. Thus, although a justification based on the rule of recognition does, of course, not necessarily imply a moral justification, the justification of the rule of recognition itself can only be based on moral reasons.

---

    1) there is general knowledge in *G* that
        a) (almost) everyone follows *R*;
        b) (almost) everyone expects (almost) everyone else to follow *R*;
        c) (almost) everyone prefers to follow *R*, provided (almost) everyone else does;
        d) (almost) everyone prefers that everyone follow some regularity, to not following any; and
    2) (almost) all members of *G* follow *R* in situations *S* precisely because the preceding conditions are fulfilled" (Bayón 1991a, 660 f.).

[13] Obviously, as a customary norm, the rule of recognition not only consists in a practice followed; here too, as with other customary norms, we must distinguish between that practice and its consideration as obligatory (that is, between the two elements traditionally distinguished in customs: *usus* and *opinio*).
[14] This point has been underscored especially by Josep Aguiló Regla (1994).
[15] On the concepts of 'operative reason' and 'auxiliary reason', cf. above, chs. I and II.

8. *How many rules of recognition? Certainty and penumbra in the rule of recognition*

Finally, we still have to treat the question of the unity of the rule of recognition and the possibility of zones of penumbra in the rule of recognition itself.

8.1. So far, we have implicitly assumed that every legal system has one and only one rule of recognition (which provides the criterion for the system's identity). That rule of recognition, however, can refer to more than one ultimate source. That is the case, for example, in the English legal system where legislation as well as precedent are sources of the law, and none of the two sources grounds that status on any criteria of validity contained in the other, respectively. According to Hart, in such cases we have a complex rule of recognition, containing more than one ultimate criterion of legal validity and a classification of these criteria "in an order of relative subordination and primacy" (Hart 1994, 101). The reason why we speak of only one rule of recognition, and not of two different ones, each of which is related to some ultimate source, is, in Hart's view, precisely that "these distinct criteria are unified by their hierarchical arrangement" (Hart 1983a, 360). However, this way of looking at things has been contested. Thus, for example, Joseph Raz writes:

> "[T]here is no reason to suppose that every legal system has just one rule of recognition. It may have more. Imagine a legal system in which no valid law makes custom or precedent a source of law, but in which, nevertheless, both custom and precedent are sources of law. It follows that the criteria for the validity of laws created by custom or precedent are determined by rules of recognition [...] But we should not assume that there is just one rule of recognition rather than two — one relating to each source of law — simply because the system must contain means of resolving conflicts between laws of the various sources. First, [...] the rule of recognition, even if it is one rule, may be incomplete, which means that the system may not include any means of resolving conflicts. Perhaps the problem has never arisen and there is no generally accepted solution to it. Secondly, there may be two or more rules of recognition that provide methods of resolving conflicts; for example, the rule imposing an obligation to apply certain customs may indicate that it is supreme, whereas the rule relating to precedent may indicate that it is subordinate." (Raz 1979, 95 f.)

Here, Raz presents two different cases: One of them is the case where there is only one rule of recognition which, however, is discovered to have gaps when one tries to determine which is the supreme criterion of legal validity and which is the subordinated one. That is possible, but, it seems, only in the context of an embryonic legal system; in any minimally developed system, it seems difficult to imagine that the question has never arisen before and has not, therefore, been given a generally accepted solution. And if it has arisen, and the judiciary or, in general, the respective legal community, instead of having created a generally accepted solution for it, is divided on a question of such central importance for the identification of valid law, then it seems that one cannot speak of *one single* legal system; rather, we would then have two systems competing with each other. The second case presented by Raz is that of two different rules of recognition, stipulating the same order of relative subordination and primacy between the two sources each one of them, respectively, refers to. But in that case, there seems to be no relevant difference to Hart's thesis of the unity of the rule of recognition: Because the assertion *a)* that a legal system has two rules of recognition each of which indicates — in a mutually consistent way — the hierarchic position of the ultimate source it refers to apparently means just the same, only formulating it differently, as the assertion *b)* that the legal system in question has one unique rule of recognition which indicates the rank order of the two ultimate sources it refers to.

8.2. Thus, the controversy over that question seems to be highly artificial. Besides, it only makes sense with respect to a legal system having more than one ultimate source. With respect to a system like the Spanish one, where there is no other ultimate source than the Constitution, the controversy cannot even be stated. More interesting for us is the question about the existence of indeterminacies, or zones of penumbra, in the rule of recognition. According to Hart, the (at least, potential) presence of zones of penumbra cannot be eliminated from any legal norm, and this also affects "the rule of recognition specifying the ultimate criteria used in the identification of the law" (Hart 1994, 123), which also "has its 'penumbrae' area as well as its 'firm', well-settled core" (Hart 1983a, 360). In the English case, that zone of penumbra affects the way in which its central pillar, that is, the doctrine of parliamentary sovereignty, is understood. Here, one must choose "between a *continuing* omnipotence in all matters not affecting the legislative competence of successive parliaments, and an unrestricted *self-embracing* omnipotence the exercise of which can only be enjoyed once" (Hart 1994, 149). Thus, the validity of a possible law of parliament that would imply that some matter is irrevocably subtracted from the future competence of the parliament itself is controversial.[16] In the Spanish case, we think

---

[16] When the United Kingdom joined the European Community, this problem, of course, acquired new relevance. But this topic by far exceeds the limits of the present study.

that the main zone of penumbra exists around the fact that by referring to the Constitution the rule of recognition refers to a source which not only claims the obligatoriness of the prescriptions issued by the authorities that source itself institutes, but also of an entire set of values and principles it contains. The possibility of conflicts between what we could call the *principle of obedience to the authorities instituted by the Constitution* — which can be seen as the most general principle, from which other, more specific principles, like those of legality, of the vinculation of ordinary courts to the interpretation of the Constitution given by the Constitutional Court, etc., derive — and the substantive principles contained in the Constitution cannot be excluded. The Constitution itself does not say what weight the ('formal') principle of obedience to the authorities instituted by the Constitution should be given, in each case, as compared to that given the 'substantive' constitutional principles (as it also does not determine the relative weight of each one of its substantive principles). This is so, not only because we are talking about a conflict between principles — that is, between legal standards whose conditions of application, as we explained earlier, are not closed, but require compliance unless in the case at hand there are other, countervailing principles that have, with respect to that case, higher weight —, but also because — just as the identity of the ultimate source(s) of a legal system is a matter of acceptance and not of prescription — the *ultimate* criteria for the interpretation of the Constitution can only be *accepted* criteria, not criteria *commanded* by the Constitution itself. And there can very well be discrepancies between the criteria accepted by different members of one and the same legal community. The protagonists probably will not even be aware of such discrepancies until they are confronted with a case where the foundation of its solution brings to light those ultimate questions of constitutional interpretation. In our view, this is precisely what happened with the judicial answer to certain cases of 'conscientious objection', mentioned above (cf. on this Atienza 1993).

8.3. Does it make sense to say that all those who accept, for example, one and the same constitution as the supreme legal source share the same rule of recognition, even if they then diverge on the interpretation of that constitution? Could one not say that different norms (rules of recognition) are underlying one and the same sentence — of the type 'One ought to obey the constitution and the norms issued or received in accordance with it'? We think that the answer to that question depends on the extent of the difference in the propositional content the different sides attribute to such a sentence. If the propositional content is basically the same for almost all members of the legal community in question, the situation can be described, following Hart, by saying that all of them share the same rule of recognition which, like all norms, besides a firm and well-established core, has a vague or open-textured periphery. Or, if you wish, one can use an alternative description (preferred by J. C. Bayón [1995]), namely: that the rule of recognition as a social norm is made up of "the area of overlap or convergence" of the rules of recognition the different members of a legal community (actually or potentially) have in mind. In any case, the fact that either (according to the first description) the rule of recognition, besides a core of certainty, also has a penumbra of doubt, or (according to the second) the rules of recognition accepted by the different members of a legal community, besides a central overlap, also have divergent peripheries, explains that ultimate questions about the identification of the law sometimes come up in controversies generated by hard cases.

CHAPTER VI
CONCLUSIONS

1. Introduction

In the course of the previous chapters, we have tried to show what basic types of legal sentences there are, and how they should be understood. We have, however, presented them in a polemic, rather than a systematic way, starting from recent controversies in legal theory about how to understand sentences expressing principles, power-conferring rules, permissions, value judgments, and the rule of recognition. This allowed us to analyze these types of sentences, and others we have found to be connected with them. But what is still missing, and what we will embark on now, is a more systematic and ordered presentation of all those types of legal sentences. We will now do this, in two phases: We will begin with a presentation and explanation of a classification of the types of legal sentences introduced earlier; and we will then subject those types to a comparative analysis, according to the three main perspectives that have guided our investigation. Tables 1 and 2 — which schematically reproduce that classification and analysis — will thus be a synthesis of the entire book.

2. A classification of legal sentences

The most general distinction one can make is that between different levels of language. It permits us to separate *legal* from *meta-legal* sentences. As we have seen, by doing this we can avoid several mistakes often made when it comes to identifying the rule of recognition, which is the only type of meta-legal sentence we have treated. Within the class of legal sentences, i. e., those belonging to some legal system, not all are (at least, not directly) of a *practical* nature, because some of them (definitions) do not — or not directly — have the function of guiding (or justifying) behaviour; instead, they identify the meaning of other sentences which do have that function in some way. Among sentences of a practical nature, we also distinguished *normative* from *evaluative* sentences, depending on whether the dominant practical function is that of guiding or of evaluating or justifying behaviour. Normative sentences, in turn, can be sentences expressing *norms*, or sentences expressing the use of powers conferred by norms. In the latter case, we speak of *normative acts*, that is, a kind of speech act where, by uttering certain words in the appropriate circumstances, one 'does' things like enacting, derogating, sentencing, etc. Thus, through normative acts, *normative changes* or *institutional results* are brought about (norms are created or cancelled, or their domain of application is restricted, or the normative status of certain individuals or groups of individuals is altered). But acts, in contrast to norms, do not last in time; they expire with performance.

Within the class of norms, one can distinguish, as is usually done, between *deontic or regulative* and *non-deontic or constitutive* norms. The former regulate behaviour, the latter determine how institutional results or normative changes are constituted or brought about. The opposition of *principles* and *rules* takes place within the class of regulative norms. As we have seen, that distinction can be drawn from three different perspectives — to which we will come back in the next phase —, but for the moment, what we are interested in is that both principles and rules can be *action norms* or *end norms*, depending on whether what is deontically modalized is an action or a state of affairs. The distinction between an *action rule* and an *end rule*, thus, runs parallel to that between *principles in the strict sense* and *policies*, in the field of principles. Since each one of these four kinds of norms can modalize behaviour with the operators *'obligatory'/'prohibited'* or *'facultative'*, we have a total of eight types of sentences expressing deontic norms, although permissive sentences — where behaviour is modalized with the operator 'facultative' — could be translated into terms of mandatory norms (and definitions).

As for non-deontic or constitutive norms, we have distinguished two main types: *power-conferring rules*, i. e., rules stipulating what one must do to produce an institutional result or a normative change, and *purely constitutive rules*, stipulating that, if a certain state of affairs obtains, then — without any need to perform an action or activity other than that which may give rise to that state of affairs — a certain institutional result or normative change is produced. Power-conferring rules can make the exercise of the conferred powers obligatory or facultative, depending on whether the result in question is modalized by a regulative norm as obligatory or as facultative; in both cases, however, performance of some specific act or course of action for bringing about the result may be optional (if there is more than one course of action that will bring about the result) or not (if there is only one such course of action).

Thus, all in all, we have distinguished 19 types of legal sentences which appear in the classification of *Table 1*. Their analysis, as to similarities and differences (basically, using the threefold scheme that distinguished between the structure of the sentences, their consideration as reasons for action, and their social function in terms of their connection with powers and interests) leads to the results contained in *Table 2*, which we will now turn to.

3. *A comparative analysis of the different types of sentences*

We will begin with the eight types of sentences expressing deontic norms.

From the structural point of view, they can all be reduced to a conditional form where the antecedent (the legal facts of the case or conditions of application) consists in a certain state of affairs, and the consequent (the legal consequence or normative solution) is constituted by an action (in the case of principles in the strict sense or of action rules) or a state of affairs (in the case of policies or end rules), modalized by mandatory operators — *1), 3), 5)* and *7)* being the four possible types of mandatory norms — or by the operator of permission. What is typical of action rules, in our view, is that they stipulate that, under certain circumstances constituting their conditions of application (which are given in closed form), a certain behaviour (which also is given in closed form) ought to, or may, be shown. End rules, in turn, are characterized by stipulating that, under the circumstances constituting their conditions of application (which are given in closed form), a certain state of affairs, which also is given in closed form, ought to, or may, be brought about. What end rules do not close is the prescribed pattern of behaviour, since they only command attainment of some state of affairs, without saying anything about the actions that are appropriate means for the purpose (which, obviously, does not mean that there may not be other normative constraints, coming from other norms of the legal order under scrutiny, concerning those actions). Principles are characterized by indicating the circumstances constituting their conditions of application in open form. The most important distinction between them is the one between principles in the strict sense and policies or program norms. The former stipulate that, except when in the case at hand there are other pertinent principles of higher weight, one must — or may — follow a certain standard of behaviour given, just as in action rules, in closed form (which implies that if there is a principle prevailing over other, possibly conflicting ones, that principle must be fully complied with). Policies, in turn, leave their conditions of application open too; but they command — or permit — to approach some state of affairs as closely as possible. Policies are similar to end rules in that they leave open the appropriate standard of behaviour for attaining the commanded state of affairs, but they differ from them in that the state of affairs also is given in open form (should be approached as closely as possible, together with other objectives whose maximization also has been ordered). Naturally, this implies that for policies, there can be different degrees of compliance.

Taking the examples given in the second column of *Table 2*, we can thus present the following 'translations' of the corresponding norms, in terms of what we have earlier said to be their 'canonical form':

1) Article 14 of the Spanish Constitution[1] (understood as a provision addressed to the law-creating and law-applying organs): "If a legal organ uses its normative powers (i. e., issues a norm in order to regulate a generic case, or applies a norm in order to solve an individual case, etc.), and if with respect to the individual or generic case in question there is an opportunity for discriminating on grounds of birth, sex, religion, opinion or any other personal or social condition, and there is no other countervailing principle of higher weight in the case at hand, then it is prohibited to that organ to discriminate on any of the grounds stated."

2) Article 20.1 a) of the Spanish Constitution:[2] "In all (generic or individual) cases where there is an opportunity for expressing or not expressing, or for diffusing or not diffusing, any idea or opinion, all those behaviours are permitted, unless there is another countervailing principle of higher weight in the case at hand." Or, translated into a mandatory principle addressed to the law-creating and law-applying organs: "In all (generic or individual) cases where there is an opportunity for the legislator, and for the public powers in general, to impose obligations or prohibitions concerning the behaviour consisting in expressing or not expressing, diffusing or not diffusing, any idea or opinion, or to hinder the performance of such behaviours, or to impose sanctions as a consequence of them, it is prohibited to the legislator, and to the public powers in general, to act in any of the stated ways, unless there is another countervailing principle of higher weight in the case at hand."

3) Article 51.1 of the Spanish Constitution[3]: "The public powers must adopt the appropriate measures for effectively protecting the security, health

---

[1] "All Spaniards are equal before the law; there shall be no discrimination on grounds of birth, race, sex, religion, opinion or any other condition or personal or social circumstance."
[2] "The following rights are recognized and protected: a) Freely to express and diffuse ideas and opinions orally, in writing, or through any other means of reproduction [...]"
[3] "The powers of the state shall guarantee the defence of consumers and users, protecting their security, health and legitimate economic interests through effective procedures."

and legitimate economic interests of consumers and users, whenever there is an opportunity for it, and to the highest degree compatible with the attainment of other ends also constitutionally ordered."

4) Article 2, paragraph 4 of EEC Directive 76/207[4] (on the application of the principle of equal treatment of men and women concerning access to employment, professional training and promotions as well as working conditions): the prohibitions of unequal treatment contained in the Directive "do not extend to measures designed to promote equal opportunities for men and women, especially in order to correct currently existing inequalities affecting the opportunities of women with respect to the matters considered in paragraph 1 of article 1 [access to employment, including promotions, professional training, working conditions, etc.]."

5) Article 28 of the Workers' Statute:[5] "If A, as an employer, has labour relations with several persons doing the same work, A must offer equal payment to all of them, irrespective of their sex."

6) Article 350 of the Spanish Civil Code:[6] "If A is the proprietor of a piece of land that is not encumbered by any easement prohibiting it, A may carry out on it whatever construction, plantation, or excavation he wishes, within the limits set by the laws and police regulations."

7) Article 103, 3 of the Spanish Civil Code:[7] "If a petition of nullity, separation or divorce has been filed, once this has been accepted and there is no agreement between the parties, the judge must, among others, take the measure he thinks necessary (in the form of guarantees, deposits, reserves, or other convenient protective measures) for securing the effectivity of the sum one spouse must pay to the other as contribution to the burdens of the marriage."

---

[4] "The present directive does not affect measures designed to promote equal opportunities for men and women, especially in order to correct currently existing inequalities affecting the opportunities of women with respect to the matters considered in paragraph 1 of article 1."
[5] "The employer is obliged to offer equal payment for equal work, concerning the basic salary as well as salary supplements, without any discrimination on grounds of sex."
[6] "The proprietor of a piece of land is the owner of its surface and of what is underneath it, and can carry out on it whatever construction, plantation, or excavation he wishes, insofar as it is not encumbered by an easement, and subject to the provisions of the laws on Mines and Water and to police regulations."
[7] "Once the petition [of nullity, separation or divorce] has been filed, and it has been judicially confirmed that there is no agreement between the spouses, the judge, in their presence, will take the following measures: ... 3. Determine the contribution of each spouse to the burdens of the marriage, including, if applicable, the *litis expensas*; establish the bases for calculating the quantities, and *fix guarantees, deposits, reserves or other convenient measures of precaution in order to secure the effectivity of the corresponding sums one spouse has to pay the other.*"

8) Article 66.1 of the General Statute of Penitentiaries:[8] "If there are groups of inmates whose treatment requires it, the prison administration may adopt measures leading to the organization, in the corresponding centers, of programs based on the principle of a therapeutic community."

In terms of reasons for action, we can distinguish between the directive and the justificatory or evaluative dimension. As for the first, that is, considered as guides for action, for those who accept them, all regulatory, mandatory norms are categorical reasons (in the Kantian sense: that is, they impose themselves on their addressee, regardless of his wishes or interests); in the typology of Joseph Raz, they are operative reasons (understood as reasons implying a practical critical attitude, that is, a critical attitude towards behaviour that corresponds to or is in conflict with that reason). According to this last classification, action rules are protected (Raz) or peremptory (Hart) rules in the strictest sense (since, when they apply, they command the exclusion of the addressee's own deliberation on what would be the best course of action to pursue as a basis for his conduct and require that he adopt as such basis the content of the rule). End rules, in turn, are peremptory reasons for bringing about some state of affairs, but they leave the choice of the means for it to the addressee's deliberation. Principles are unprotected, non-peremptory reasons (since their prevalence — in the case of principles in the strict sense — or the relationships between different, interrelated objectives — in the case of policies — requires the addressee's deliberation).

As for permissive regulatory norms, seen from this perspective, they express that for the addressee there are no specific reasons for or against the performance of the action in question (either saying that there actually are no such reasons, or cancelling those that may have existed before), and that for third parties there is a reason for not preventing or hindering the addressee from performing or not performing the action in question; in the case of permissive principles (principles in the strict sense or policies), that reason is a non-peremptory reason, whereas for permissive action or end rules, it is a peremptory reason. In the case of rules, permissive provisions (regulating natural behaviour, that is, behaviour not consisting in the exercise of normative powers), may be explicated entirely in terms of the clarification (through definition), exception, derogation or indirect formulation of mandatory rules (prohibiting

---

[8] "For certain groups of inmates whose treatment requires it, in the corresponding centers programs based on the principle of therapeutic community may be organized."

lower-ranking authorities to introduce prohibitions, impediments or sanctions). In the case of principles, permissive provisions whose explicit addressee is the general public (basically, constitutional liberties) can be regarded as *prima facie* prohibitions for the legislator and, generally, for all lower authorities, to interfere in certain spheres reserved to autonomy (principles in the strict sense) and as commands to adopt measures for making that autonomy maximally effective in the respective sphere (policies). As in the case of the example contained in *Table 2*, there may also be permissive policies whose addressees are public powers. In order to maximize the attainment of some end (in our case, an equal standing of men and women as social groups) such policies state an exception to some general prohibition addressed to those same public powers, or say that it is not applicable in the area in question (in our case, the prohibition of giving preferential treatment to some person on the ground that she does, or does not, belong to one of those groups).

If we now proceed to look at deontic or regulatory norms from the justificatory or evaluative dimension of reasons for action, the distinction between principles in the strict sense and policies is as follows: Principles in the strict sense are reasons of correctness that presuppose ultimate values, whereas in the case of policies we have utilitarian reasons presupposing values of that same kind. This implies that the second type of reasons can and should be evaluated — and possibly overruled — by reasons of correctness, whereas the contrary cannot legitimately occur. As for rules of any of the four types we have distinguished, they can incorporate ultimate values as well as utilitarian or simply instrumental values (or some combination thereof).

Finally, from the perspective of the social function fulfilled by regulatory norms, with respect to social power and interests, the difference between principles and rules can be expressed as follows: action rules do not require one to weigh and balance the interests and values in question; end rules require such balancing only with respect to means (they concede, so to speak, a certain degree of discretion), but not to ends; principles in the strict sense require balancing when they are applied, but do not give the organ of application any power of discretion; and policies require balancing, and concede power of discretion with respect to means and ends. Besides, with respect to principles, principles in the strict sense mainly have a negative or constraining function in the pursuit of interests, whereas policies positively promote the attainment of objectives constituting social interests. On the other hand, in the case of rules, all of them constrain the pursuit of individual or social interests, guaranteeing a sphere of non-interference (permissive rules), or generating mutual constraints through the stipulation of positive and negative duties (mandatory rules).

Let us now look at power-conferring rules. From the structural point of view, those rules link the production of an institutional result (or, what amounts to the same, a normative change) to the existence of a certain state of affairs, together with the performance of an action by some subject. Their canonical form is: 'If state of affairs X obtains, and Z performs action Y, then institutional result (normative change) R is produced'. In our opinion, such norms are anankastic-constitutive rules. They are constitutive because the rule itself creates the possibility for bringing about the institutional result (which does not exist independently of the rule); and they are anankastic because on the basis of such rules their addressees can construct institutional technical rules (just as on the basis of anankastic natural propositions one can construct natural technical rules).

In *Table 2*, we analyze only power-conferring rules whose exercise is obligatory or whose exercise is facultative; we thus do not consider the possible distinction between whether or not element Y — the action of the antecedent — is optional. The two examples we give of those two main types of power-conferring rules would be expressed as follows in their canonical form:

9) and *10)* Article 160 of the Spanish Constitution:[9] "If circumstances X obtain (that A is a member of the Constitutional Court and the Court's plenum proposes him for President) and if Z (the King) performs action Y (appointing A), then normative result R (that A becomes President of the Constitutional Court) is produced."

11) and *12)* Article 1254 of the Spanish Civil Code:[10] "If state of affairs X obtains (there is at least one person with full legal capacity) and those persons perform action Y (agreeing to commit themselves to give something or to perform some service to some other person or persons), then normative result R (there is a contract) is produced."

If we now consider power-conferring rules in terms of reasons for action, they are — in their directive aspect — auxiliary (Raz) or hypothetical (in the Kantian classification of imperatives) reasons. They indicate how one must act

---

[9] "The president of the Constitutional Court is appointed from among its members by the King, on a proposal by the plenum of the Court itself, for a term of three years."
[10] "There is a contract from the moment that one or more persons agree to incur the obligation of giving something or performing some service to some other person or persons."

in order to bring about a certain institutional result or normative change. The operative reason to bring this about may be either a mandatory norm making it obligatory (as is the case, for example, with judicial power: the judge is obliged to produce the institutional result of a 'judicial decision') or a wish or interest of the agent, when the decision to produce or not produce the result is not subject to normative constraints (as is the case, for example, of private contractual powers). In the first case, the power-conferring rule is an assertoric hypothetical, in the second a problematic hypothetical reason. When the bringing-about of the normative result is prohibited by a mandatory norm, it means that in such a case the power-conferring rule cannot be legitimately used as a reason for action.

Seen as reasons for action, but now from the evaluative or justificatory perspective, element Y always represents a purely instrumental value, whereas R — the normative result — can be an ultimate, a utilitarian, or simply an instrumental value in the case of a power-conferring rule of obligatory exercise; when the exercise is facultative, however, R is indifferent (from the point of view of the legal order, of course, not from that of someone using the rule).

As for their social function, power-conferring rules of any of the two kinds, through the modification of one's own or others' normative status, indirectly promote the pursuit of one's own or others' interests. However, that interests are thus affected does not depend — in the case of rules of obligatory exercise — on the wishes or interests (but on the action) of the addressee to whom power is conferred, in contrast to what happens in the case of power-conferring rules of facultative exercise.

Besides power-conferring (anankastic-constitutive) rules, the legal system also contains purely constitutive rules which also determine the production of institutional results or normative changes. These are rules whose canonical form is: 'If state of affairs X obtains, then institutional result (or normative change) R is produced.' As can be seen from the example we have given for this kind of rules, the normative change (the transmission of rights of succession) is produced without any need for the person affected by the change to perform an action:

13) Article 657 of the Spanish Civil Code:[11] "When a person dies, her rights of succession are passed on at the moment of death."

---

[11] " A person's rights of succession are passed on at the moment of her death."

Since they do not refer to any action, such rules are not really reasons for action (for bringing about R), but they may provide auxiliary reasons for bringing about the state of affairs constituting the antecedent of the rule (provided the bringing-about of that state of affairs is permitted and is under the control of the agent). In its evaluative dimension, from the point of view of the legal order, R is an indifferent state of affairs. Purely constitutive rules affect interests, but (in contrast to what happens with power-conferring rules) that does not depend on the performance of a normative act by anyone.

Normative acts can be analyzed, we think, by distinguishing their locutionary (what is said), their illocutionary (what is done by saying it), and their perlocutionary (what are the effects of doing it) aspects. These three levels have a certain similarity with the three perspectives we have distinguished in norms. In contrast to norm-expressing sentences, those expressing the use of normative powers — whose uttering, in the appropriate conditions, constitutes the use of a normative power — have no conditional structure, as can be seen in the following example:

14) Derogating provision of the Spanish Criminal Code of 1995, 1.c): "Derogated are: [...] c) Law 16/1970 of August 4, on Dangerousness and Social Rehabilitiation, with later modifications and supplementary provisions."

Such sentences do not express any reasons for action, but speech acts that are instances of the use of normative powers; by introducing, eliminating, using or applying norms of the preceding types, they affect in different ways individual and social interests.

As for evaluative sentences, we have considered them, from an internal perspective, as the justificatory element contained in the preceding types of norms. What characterizes an evaluative sentence is that the emphasis is on the justificatory aspect; but we have held that they can be translated without any loss of meaning into normative sentences of one or the other of the types we have distinguished. Thus, in the case of article 1.1 of the Spanish Constitution:

15) Article 1.1.: "Spain is constituted as a social and democratic State under the rule of law, *declaring as the highest values of its legal order liberty, justice, equality and political pluralism*"

an equivalent formulation would be:

15') "In the Spanish legal order, *acts of norm-creation and norm-application must implement the principles of liberty, justice, equality and political pluralism*".

Definitions are sentences without any practical dimension, that is, they are neither guides nor criteria of evaluation for any behaviour. Their structure is not conditional, but has the canonical form: "'...' means '...'", where the *definiendum* and the *definiens* are terms or concepts, not actions or states of affairs. Thus:

16) Article 660 of the Spanish Civil Code:[12] "Heir" means "person entitled to inherit an estate as a whole", and "legatee" means "person entitled to a partial inheritance".

Their function is to identify norms, understood as the meaning of normative sentences, by elucidating the meaning given to certain expressions used in such norms. Besides, legislative definitions not only fulfil the (explicatory) function of clarifying language, by reducing ambiguities and vagueness, but also (if you wish, by thus specifying the terms) that of restricting — or extending — the 'semantic power' of judges and legal doctrine.

Finally, the rule of recognition, constituting a guide and criterion of evaluation for behaviour and decisions, on the one hand, and providing a theoretical criterion for identifying legal norms, on the other, reflects on the meta-legal level the three broad types of sentences distinguished on the level of the language of the law: the normative, the evaluative and the conceptual (defining) types. In *Table 2*, however, we distinguish only two of its forms: as a practical sentence, and as a theoretical or conceptual sentence:

17) and *18):* "The norms contained in the Constitution of 1978 and the norms issued or received in accordance with it, ought to be obeyed"; "behaviour and decisions ought to be evaluated exclusively on the basis of the norms contained in the Constitution of 1978 and the norms issued or received in accordance with it".

19) "Norms of the Spanish legal order are those contained in the Constitution of 1978 and in the sources recognized or instituted by it."

---

[12] "Heir is the person entitled to inherit an estate as a whole, legatee the person entitled to a partial inheritance."

Acceptance of the rule of recognition as a guide and criterion of evaluation amounts to acceptance of the exclusionary claim of the law, that is, the assumption that when its norms apply, one ought to behave and judge on the basis of them, excluding any reasons other than those norms (or the reasons those norms possibly refer to) as a basis of action and evaluation. As a theoretical criterion of identification, the rule of recognition gives unity to the legal order, by delimiting the domain of legal norms: they are those the rule of recognition itself directly or indirectly refers to.

Seen under its normative aspect, the rule of recognition is a mandatory norm which, in its directive dimension, works as a peremptory reason in favor of the legal system as a whole and which, in its justificatory dimension, provides the ultimate criterion of legal evaluation. Thus, the rule of recognition — as guide of behaviour and criterion of evaluation — is the ultimate legal norm; that is why its acceptance cannot be justified on the basis of any reason given by the law, but only on the basis of impartial and autonomous reasons — reasons, that is, whose value is in their intrinsic merits — because to accept the rule of recognition implies accepting the normative claims of the law which, obviously, are not restricted only to those who accept them, but extend to all persons under the conditions of application stipulated by the legal norms themselves. But to speak of autonomous and impartial reasons is the same as to speak of moral reasons.

Table 1. Classification of Sentences

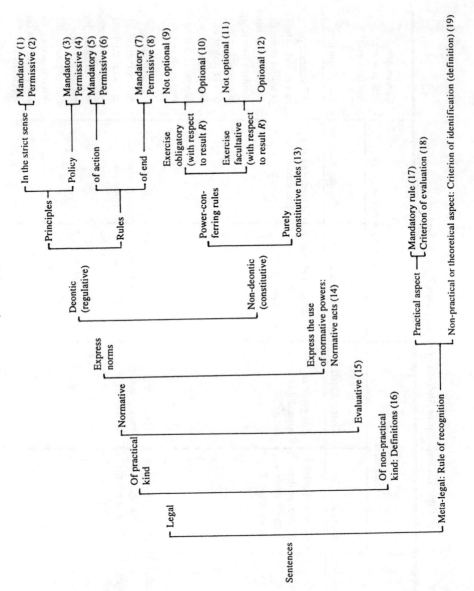

Table 2. Comparative Analysis of the Different Types of Sentences

Sentences	Example	Structure		Reasons for action		Connection with powers and interests*
		Canonical form	Characteristics	Directive function	Justific. function	Social function
Mandatory principles in the strict sense (1)	art. 14 Spanish Constit.	If state of affairs X obtains, then Z ought to perform action Y	X is given in open, Y in closed form	– Categorical reason – Non-peremptory operative reason	Ultimate value (reasons of correctness)	– Constrain pursuit of individual or social interests – Their application requires balancing, but they do not confer power of discretion on adressees nor on applicators
Permissive principles in the strict sense (2)	art. 20, 1.a) Spanish Constit.	If state of affairs X obtains, then Z may perform action Y	X is given in open, Y in closed form	– Categorical reason – Non-peremptory operative reason for not hindering or preventing Z from doing or not doing Y – No reasons for Z concerning Y	Ultimate value (esp. value of autonomy)	– Constrain pursuit of individual or social interests – Their application requires balancing, but they do not confer power of discretion on adressees nor on applicators

Mandatory policies (3)	art. 51.1 Spanish Constit.	If state of affairs X obtains, then Z ought to try to reach end (state of affairs) F	X and F are given in open form	– Categorical reason – Non-peremptory operative reason	Utilitarian value (teleological reasons controlled by reasons of correctness)	– Promote satisfaction of social interests – Their application requires balancing; they confer power of discretion on addressees (concerning means and relations between ends)
Permissive policies (4)	art. 2, sect. 4 EEC Directive 76/207	If state of affairs X obtains, then Z may try to reach end (state of affairs) F	X and F are given in open form	– Categorical reason – Non-peremptory operative reason for not hindering or preventing Z from bringing or not bringing about F – No reasons for Z concerning Y	Utilitarian value (teleological reasons controlled by reasons of correctness)	– Promote satisfaction of social interests – Their application requires balancing; they confer power of discretion on addressees (concerning means and ends)
Mandatory action rules (5)	art. 28 Spanish Workers' Statute	If state of affairs X obtains, then Z ought to perform action Y	X and Y are given in closed form	– Categorical reason – Peremptory operative reason in the strict sense	Ultimate, utilitarian or simply instrumental value	– Constrain pursuit of individual or social interests, guaranteeing a sphere of non-intervention – Do not require balancing of given interests

Table 2. (continued)

Sentences	Structure			Reasons for action		Connection with powers and interests*
	Example	Canonical form	Characteristics	Directive function	Justific. function	Social function
Permissive action rules (6)	art. 350, Spanish Civil Code	If state of affairs X obtains, then Z may perform action Y	X and Y are given in closed form	– Categorical reason – Peremptory operative reason for not interfering in Z's doing Y – No reasons for Z concerning Y	Ultimate, utilitarian, or simply instrumental value	– Constrain pursuit of individual or social interests, guaranteeing a sphere of non-interference – Do not require balancing of given interests
Mandatory end rules (7)	art. 103, 3rd Spanish Civil Code	If state of affairs X obtains, then Z ought to reach end (state of affairs) F	X and F are given in closed form	– Categorical reason – Peremptory operative reason for reaching F; deliberation on means left to Z	Ultimate, utilitarian, or simply instrumental value	– Constrain pursuit of individual or social interests, imposing positive and negative duties, thus creating mutual constraints – Give power of discretion (concerning means)

Permissive end rules (8)	art. 66.1 Spanish General Statute of Prisons	If state of affairs X obtains, then Z may try to reach end (state of affairs) F	X and F are given in closed form	– Categorical reason – Peremptory operative reason for not interfering in Z's bringing about F – No reasons for Z concerning F	Ultimate, utilitarian, or simply instrumental value	– Constrain pursuit of individual or social interests, guaranteeing a sphere of non-interference – Give power of discretion (concerning means)
Power-conferring rules of obligatory exercise (9) and (10)	art. 160 Spanish Constit.	If state of affairs X obtains and Z performs Y, then institutional result (normative change) R is produced	Anankastic-constitutive norm from which the addressee can construct institutional technical rules	– Auxiliary reason – Hypothetical assertorial reason: the operative reason for producing R (the normative change) is a mandatory norm	– Y is a purely instrumental value – R is an ultimate, utilitarian, or simply instrumental value	– Indirectly enable one to pursue own or others' interests (through modification of own or others' normative status (R)) – Effect on interests does not depend on wishes (but on actions) of rule's addressees

Table 2. (continued)

Sentences	Example	Structure		Reasons for action			Connection with powers and interests*
		Canonical form	Characteristics	Directive function	Justific. function		Social function
Power-conferring rules of facultative exercise (11) and (12)	art. 1254 Spanish Civil Code	If state of affairs X obtains and Z performs Y, then institutional result (normative change) R is produced	Anankastic-constitutive norm from which the addressee can construct institutional technical rules	– Auxiliary reason – Hypothetical problematical reason: the operative reason for producing R is the agent's wish or interest	– Y is a purely instrumental value – R is an indifferent value (for the legal order)		– Indirectly enable one to pursue own or others' interests (through modification of own or others' normative status [R]) – Effect on interests depends on wishes and interests of rule's addressees
Purely-constitutive rules (13)	art. 657 Spanish Civil Code	If state of affairs X obtains, then institutional result (normative change) R is produced	No action needs to be performed for the normative change to come about	– No direct reasons for action (for producing R) – Can be auxiliary or hypothetical reasons for producing the state of affairs constituting the antecedent (provided production of it is permitted and under agent's control)	– R is an indifferent state of affairs (for the legal order)		– Indirectly enable one to pursue own or others' interests (through constitution of R) – Effect on interests does not depend on wishes, interests or action of person affected by norm. change

Normative acts (14)	Derogating provision, Spanish Criminal Code of 1995, 1.C.	Don't have conditional structure	Correspond to the locutionary aspect: 'Law L is derogated.'	– They are not reasons for action, but actions performed by making use of a normative power – Correspond to the illocutionary aspect, i. e. the force of the sentence	– Correspond to the perlocutionary aspect, i. e. the effects: by introducing, cancelling, or applying norms of any of the previous kinds, individual or social interests are affected in various ways
Values (14)	art. 1.1 Spanish Constit.			– Evaluative sentences are considered - from an internal perspective - the justificatory element of norms – Evaluative 'sentences' can be translated into normative 'sentences'	
Definitions (16)	art. 360 Spanish Civil Code	"..." means ...	"..." and ... are terms or concepts (not actions or states of affairs)	– They are not reasons for action – Their function is to identify norms by elucidating the meaning of words	– Reduce or extend the 'semantic power' of law-applying organs
Rule of recognition (17) and (18)	'The Span. Constit. of 1978 ought to be obeyed'	Independent norms and norms issued or received in accordance with them ought to be obeyed	Mandatory customary rule	– Categorical reason – Peremptory operative reason for the legal system as a whole – Provides ultimate criteria of legal, not moral justification	Ultimate reason from the point of view of the legal system, not from a moral point of view	– Reflects acceptance of the normative claim of the law: its capacity to regulate behaviour exclusively, but not unrestrictedly

Table 2. (continued)

Sentences	Structure			Reasons for action		Connection with powers and interests*
	Canonical form	Example	Characteristics	Directive function	Justific. function	Social function
Rule of recognition (19)	Legal norms are independent norms and norms issued or received in accordance with them	'Spanish legal norms are the norms of the Spanish Constit. and those issued or received in accordance with it'	Definition	– Is no reason for action (has no practical function) – Serves to identify (recognize) the norms of the system		Enables one to describe a legal system as having unity (reflects the unitarian character of the law)

* The analysis in terms of powers and interests is carried out from the perspective of the primary systems, not from that of the organs of application. In some cases, we have used examples of norms addressed to the judges; obviously, the interests alluded to are not those of the judges themselves, but of those affected by those norms. Besides, norms addressed to the organs of application certainly restrict their power, but this is a normative power internal to the legal order, not social power (linked to interests external to the order). Definitions – including the rule of recognition, understood as a conceptual criterion – do not affect interests and power relations directly, but through the norms whose content they help stipulate. Therefore, in the Table, in those cases we have referred to their effect on organs of application and legal doctrine.

BIBLIOGRAPHY

Aarnio, Aulis (1987): The Rational as Reasonable, Dordrecht: Reidel.

Aguiló Regla, Josep (1990): Sobre definiciones y normas, Doxa (Alicante) 8.

Aguiló Regla, Josep (1993): Sobre la derogación. Ensayo de dinámica jurídica, México: Fontamara.

Aguiló Regla, Josep (1994): Buenos y malos. Sobre el valor epistémico de las actitudes morales y de las prudenciales, in: Doxa (Alicante) 15-16.

Alchourrón, Carlos E. and Eugenio Bulygin (1971): Normative Systems, Vienna and New York: Springer.

Alchourrón, Carlos E. and Eugenio Bulygin (1974): Introducción a la metodología de las ciencias jurídicas y sociales, Buenos Aires: Astrea.

Alchourrón, Carlos E. and Eugenio Bulygin (1981): The Expressive Conception of Norms, in: R. Hilpinen (ed.), New Studies in Deontic Logic, Dordrecht et al.: Reidel, 95-124.

Alchourrón, Carlos E. and Eugenio Bulygin (1984): Permission and Permissive Norms, in: W. Krawietz et al. (eds.), Theorie der Normen, Berlin: Duncker & Humblot, 349-371.

Alchourrón, Carlos E. and Eugenio Bulygin (1985): Libertad y autoridad normativa, in: Boletín de la Asociación Argentina de Filosofía del Derecho (La Plata) vol. 3 no. 26.

Alchourrón, Carlos E. and Eugenio Bulygin (1991): Análisis lógico y Derecho, with an introduction by G. H. von Wright, Madrid: Centro de Estudios Constitucionales.

Alchourrón, Carlos E. and Eugenio Bulygin (1991a): Definiciones y normas (1983), in: Alchourrón/Bulygin 1991, 439-463.

Alexy, Robert (1986): Theorie der Grundrechte, Frankfurt a. M.: Suhrkamp.

Alexy, Robert (1988): Sistema jurídico, principios jurídicos y razón práctica, Doxa (Alicante) 5.

Atienza, Manuel (1986): La analogía en el Derecho. Ensayo de análisis de un razonamiento jurídico, Madrid: Civitas.

Atienza, Manuel (1993): Tras la justicia. Una introducción al Derecho y al razonamiento jurídico, Barcelona: Ariel.

Atienza, Manuel (1995): Sobre el control de la discrecionalidad administrativa. Comentarios a una polémica, in: Revista Española de Derecho Administrativo 85.

Atienza, Manuel, and Juan Ruiz Manero (1991): Sobre principios y reglas, in: Doxa 10.

Atienza, Manuel, and Juan Ruiz Manero (1992): Objeciones de principio. Respuesta a Aleksander Peczenik y Luis Prieto Sanchís, in: Doxa (Alicante) 12.

Atienza, Manuel, and Juan Ruiz Manero (1993): Tre approcci ai principi di diritto, in: Analisi e diritto 1993, Torino: Giappichelli.

Atienza, Manuel, and Juan Ruiz Manero (1994a): Sulle regole que conferiscono poteri, in: Analisi e diritto 1994, Torino: Giappichelli.

Atienza, Manuel, and Juan Ruiz Manero (1994b): Sobre permisos en el Derecho, in: Doxa (Alicante) 15-16.

Azzoni, Gianpaolo (1988): Il concetto di condizione nella tipologia delle regole, Padua: Cedam.

Bayón Mohino, Juan Carlos (1991a): La normatividad del Derecho. Deber jurídico y razones para la acción, Madrid: Centro de Estudios Constitucionales [English version: The Normativity of Law. Legal Duty and Reasons for Action, Dordrecht: Kluwer, in preparation].

Bayón Mohino, Juan Carlos (1991b): Razones y reglas. Sobre el concepto de 'razón excluyente' de Joseph Raz, in: Doxa (Alicante) 10.

Bayón Mohino, Juan Carlos (1995): Participantes, observadores e identificación del Derecho, paper presented at the Spanish-Italian Seminar of Legal Theory, Imperia, Italy.

Bobbio, Norberto (1966): Principi generali del diritto, in: Novissimo Digesto Italiano, XIII, Torino: UTET.

Bulygin, Eugenio (1976): Sobre la regla de reconocimiento, in: idem et al., Derecho, filosofía y lenguaje. Homenaje a Ambrosio L. Gioja, Buenos Aires: Astrea.

Bulygin, Eugenio (1991): Sobre las normas de competencia, in: Alchourrón/ Bulygin 1991, 485-498.

Bulygin, Eugenio (1991a): Algunas consideraciones sobre los sistemas jurídicos, in: Doxa (Alicante) 9.

Bulygin, Eugenio (1991b): Regla de reconocimiento: ¿norma de obligación o criterio conceptual? Réplica a Juan Ruiz Manero, in: Doxa (Alicante) 9.

Bulygin, Eugenio (1992): On Norms of Competence, in: Law and Philosophy 11, 201-216.

Caracciolo, Ricardo (1988): El sistema jurídico. Problemas actuales, Madrid: Centro de Estudios Constitucionales.

Caracciolo, Ricardo (1991): Sistema jurídico y regla de reconocimiento, in: Doxa (Alicante) 9.

Caracciolo, Ricardo (1995): Due tipi di potere normativo, in: Analisi e diritto 1995, Torino: Giappichelli.

Carrió, Genaro R. (1986): Principios jurídicos y positivismo jurídico, in: idem, Notas sobre Derecho y lenguaje, 3d ed. Buenos Aires: Abeledo Perrot.

Cobo del Rosal, Manuel and Tomás S. Vives Antón (1990): Derecho Penal. Parte general, Valencia: Tirant lo Blanc.

Conte, Amedeo (1985a): Materiali per una tipologia delle regole. Materiali per una storia della cultura giuridica 15, 345 ff.

Conte, Amedeo (1985b): Phénoménologie du langage déontique, in: G. Kalinowski and S. Filippo (eds.), Les fondements logiques de la pensée normative, Rome: Università Gregoriana.

De Otto, Ignacio (1989): Derecho constitucional. Sistema de fuentes, 2nd ed. 1st reprint, Barcelona: Ariel.

Del Vecchio, Giorgio (1958): Sui principi generali del diritto, in: idem, Studi sul diritto, vol. I, Milan.

Díez-Picazo, Luis, and Antonio Gullón (1989): Sistema de Derecho Civil, vol. I, 7th ed. Madrid: Tecnos.

Dworkin, Ronald (1978): Taking Rights Seriously, London: Duckworth.

Dworkin, Ronald (1986): Law's Empire, London: Fontana.

Echave, Delia Teresa, María Eugenia Urquijo and Ricardo Guibourg (1980): Lógica, proposición y norma, Buenos Aires: Astrea.

Esser, Joseph (1956): Grundsatz und Norm in der richterlichen Fortbildung des Privatrechts, Tübingen: J. C. B. Mohr.

Fishkin, James S. (1982): The Limits of Obligation, New Haven and London: Yale University Press.

Fishkin, James S. (1986): Las fronteras de la obligación, in: Doxa (Alicante) 3.

García de Enterría, Eduardo (1963): Reflexiones sobre la Ley y los principios generales del Derecho en el Derecho administrativo, in: Revista de Administración Pública 40 [reprinted in: idem, Reflexiones sobre la Ley y los principios generales del Derecho, Madrid: Civitas 1984).

García de Enterría, Eduardo (1988): La Constitución como norma y el Tribunal Constitucional (1963), reprint of the 3rd ed. Madrid: Civitas.

González Lagier, Daniel (1993): Clasificar acciones. Sobre la crítica de Raz a las reglas constitutivas de Searle, Doxa (Alicante) 13.

González Lagier, Daniel (1995): Acción y norma en G. H. von Wright, Madrid: Centro de Estudios Constitucionales.

Greenawalt, Kent (1988): Hart's Rule of Recognition and the United States, in: Ratio Juris 1:1.

Guastini, Riccardo (1983): Teorie delle regole costitutive, in: RIFD 60.

Guastini, Riccardo (1985): Questioni di tecnica legislativa, in: Le regioni 2-3.

Guastini, Riccardo (1990): Reglas constitutivas y gran división, in: Cuadernos del Instituto de Investigaciones Jurídicas (México) 14.

Guastini, Riccardo (1990a): Principi di diritto, in: Dalle fonti alle norme, Turino: Giappichelli.

Guastini, Riccardo (1993): Le fonti del diritto e l'interpretazione, Milan: Giuffrè.

Guibourg, Ricardo (1993): Hart, Bulygin y Ruiz Manero. Tres enfoques para un modelo, in: Doxa (Alicante) 14.

Guibourg, Ricardo and Daniel Mendonça (1995): Permesso, garanzie, e libertà, in: Analisi e diritto 1995, Torino: Giappichelli.

Hare, Richard M. (1952): The Language of Morals, Oxford: Clarendon.

Hart, H. L. A. (1982a): Commands and Authoritative Legal Reasons, in: Essays on Bentham. Jurisprudence and Legal Theory, Oxford: Clarendon.

Hart, H. L. A. (1982b): Legal Powers, in: Essays on Bentham. Jurisprudence and Legal Theory, Oxford: Clarendon.

Hart, H. L. A. (1983a): Review of Lon L. Fuller, *The Morality of Law* (1964), in: Harvard Law Review 1281 (1965), reprinted in idem, Essays in Jurisprudence and Philosophy, Oxford: Clarendon.

Hart, H. L. A. (1983b): Kelsen Visited, in: idem, Essays in Jurisprudence and Philosophy, Oxford: Clarendon.

Hart, H. L. A. (1994): The Concept of Law (1961), 2nd ed., Oxford: Clarendon.

Hernández Marín, Rafael (1984): El Derecho como dogma, Madrid: Tecnos.

Hernández Marín, Rafael (1993): Double Pairs, in: Ratio Juris 6:3.

Jori, Mario (1995): Definizioni giuridiche e pragmatica, in: Analisi e diritto.

Kelsen, Hans (1967): The Pure Theory of Law, translated by Max Knight, University of California Press [German original: Reine Rechtslehre, 2nd ed. Vienna: Deuticke 1960].

Kelsen, Hans (1973): Derogation, in: Essays in Legal and Moral Philosophy, ed. and introd. by O. Weinberger, Dordrecht: Reidel.

Laporta, Franciso J. (1984): Norma básica, Constitución y decisión por mayorías, in: Revista de las Cortes Generales I.

Laporta, Francisco J. (1987): Sobre el concepto de derechos humanos, in: Doxa (Alicante) 4.

Luhmann, Niklas (1974): Rechtssystem und Rechtsdogmatik, Stuttgart et al.: Kohlhammer.

Lukes, Steven (1974): Power. A Radical View, Houndmills and London: Macmillan.

MacCormick, Neil (1981): H. L. A. Hart, London: Edward Arnold.

MacCormick, Neil (1986): Law as Institutional Fact, in: Neil MacCormick and Ota Weinberger (eds.), An Institutional Theory of Law. New Approaches to Legal Positivism, Dordrecht: Reidel.

MacCormick, Neil (1993): Powers and power-conferring norms, paper presented at the 5th Kelsen Seminar, Siena.

Mendonça, Daniel, José Juan Moreso and Pablo Navarro (1995): Intorno alle norme di competenza, in: Analisi e diritto 1995, Torino: Giappichelli.

Mir Puig, Santiago (1976): Introducción a las bases del Derecho penal, Barcelona: Bosch.

Mir Puig, Santiago (1990): Derecho Penal. Parte general, 3rd ed., Barcelona: PPV.

Moreso, José Juan, Pablo E. Navarro and Cristina Redondo (1992): Argumentación jurídica, lógica y decisión judicial, in: Doxa (Alicante) 11.

Nino, Carlos S. (1980): Introducción al análisis del Derecho, Buenos Aires: Astrea.

Nino, Carlos S. (1985): La validez del Derecho, Buenos Aires: Astrea.

Nino, Carlos S. (1985a): ¿Son prescripciones los juicios de valor?, in: idem, La validez del Derecho, Buenos Aires: Astrea.

Nino, Carlos S. (1985b): Normas jurídicas y razones para actuar, in: idem, La validez del Derecho, Buenos Aires: Astrea.

Nino, Carlos S. (1992): Fundamentos de Derecho constitucional, Buenos Aires: Astrea.

Nino, Carlos S. (1994): Derecho, moral y política. Una revisión de la teoría general del Derecho, Barcelona: Ariel.

Peces-Barba Martínez, Gregorio (1984): Los valores superiores, Madrid: Tecnos.

Peczenik, Aleksander (1989): On Law and Reason, Dordrecht et al.: Kluwer.

Peczenik, Aleksander (1990): Dimensiones morales del Derecho, in: Doxa (Alicante) 8.

Peczenik, Aleksander (1992): Los principios jurídicos según Manuel Atienza y Juan Ruiz Manero, in: Doxa (Alicante) 12.

Prieto Sanchís, Luis (1985): Teoría del Derecho y filosofía política en Ronald Dworkin, in: Revista Española de Derecho Constitucional 14.

Prieto Sanchís, Luis (1991): Notas sobre la interpretación constitucional, in: Revista del Centro de Estudios Constitucionales (Madrid) 9.

Prieto Sanchís, Luis (1992): Sobre principios y normas. Problemas del razonamiento jurídico, Madrid: Centro de Estudios Constitucionales.

Prieto Sanchís, Luis (1993): Dúplica a los profesores Manuel Atienza y Juan Ruiz Manero, in: Doxa (Alicante) 13.

Rawls, John (1971): A Theory of Justice, Cambridge, Mass.: Harvard.

Raz, Joseph (1979): The Authority of Law, Oxford: Clarendon.

Raz, Joseph (1980): Postscript. Sources, Normativity, and Individuation, in: The Concept of a Legal System. An Introduction to the Theory of Legal Systems, 2nd ed. Oxford: Clarendon.

Raz, Joseph (1986): The Morality of Freedom, Oxford: Clarendon.

Raz, Joseph (1990): Practical Reason and Norms (1975), Princeton, NJ: Princeton University Press.

Regan, Donald (1989): Authority and Value: Reflections on Raz's *Morality of Freedom*, in: Southern California Law Review 62.

Rescher, Nicholas (1969): Introduction to Value Theory, Englewood Cliffs, NJ: Prentice-Hall.

Robles, Gregorio (1984): Las reglas del Derecho y las reglas de los juegos. Ensayo de teoría analítica del Derecho, Palma de Mallorca: Universidad de Palma de Mallorca.

Robles, Gregorio (1986): La comparación entre el Derecho y los juegos, in: Anuario de Filosofía del Derecho (Madrid).

Rodilla, Miguel Angel (1986): El Derecho y los juegos. Utilidad y límites de una analogía, in: Anuario de Filosofía del Derecho (Madrid).

Ross, Alf (1968): Directives and Norms, London: Routledge & Kegan Paul.

Rubio Llorente, Francisco (1995): Prólogo, in: F. Rubio Llorente et al., Derechos fundamentales y principios constitucionales, Barcelona: Ariel.

Ruiz Manero, Juan (1990): Jurisdicción y Normas, Madrid: Centro de Estudios Constitucionales.

Ruiz Manero, Juan (1991): Normas independientes, criterios conceptuales y trucos verbales. Respuesta a Eugenio Bulygin, in: Doxa (Alicante) 9.

Ruiz Manero, Juan (1992): Respuesta a Luis Martínez Roldán, in: Anuario de Filosofía del Derecho (Madrid).

Ruiz Manero, Juan (1994): On the Alternative Tacit Clause, in: Letizia Gianformaggio and Stanley L. Paulson (eds.), Cognition and Interpretation of Law, Turino: Giappichelli.

Ruiz Miguel, Alfonso (1988): El principio de jerarquía normativa, in: Revista Española de Derecho Constitucional 8:24.

Schauer, Frederick (1991): Playing by the Rules. A Philosophical Examination of Rule-Based Decision-Making in Law and in Life, Oxford: Clarendon.

Silva Sánchez, José María (1992): Aproximación al Derecho penal contemporáneo, Barcelona: Bosch.

Summers, Robert S. (1978): Two Types of Substantive Reasons, in: Cornell Law Review 63.

Vernengo, Roberto J. (1990): Los derechos humanos como razones morales justificatorias, in: Doxa (Alicante) 7.

Vives Antón, Tomás S. (1979): Concepto, método y fuentes del Derecho penal, Valencia: unpublished manuscript.

von Wright, Georg Henrik (1963): Norm and Action. A Logical Enquiry, London: Routledge & Kegan Paul and New York: Humanities Press.

von Wright, Georg Henrik (1968): Deontic Logic and the Theory of Conditions, in: Crítica (Mexico-City) vol. II no. 6.

von Wright, Georg Henrik (1969): On the Logic and Ontology of Norms, in: J. W. Davis (ed.), Philosophical Logic, Dordrecht: Reidel.

INDEX OF NAMES

Aarnio, A. 25
Ackerman, B. 128
Aguiló Regla, J. 56, 63, 65, 157
Alchourrón, Carlos E. 5 ff., 17, 27, 29 ff., 46 ff., 53, 55 f., 62 ff., 67, 84 f., 89, 95, 99 ff., 104, 111 f., 153
Alexy, R. 9, 11, 28, 37 f., 41 f.
Atienza, M. 13, 28, 32, 38, 42, 76, 116, 160
Azzoni, G. 61 f., 66

Bayón Mohino, J. C. 16, 33, 36, 112, 128 f., 156 f., 161
Bobbio, N. 3
Bulygin, E. 5 ff., 17, 27, 29 f., 46 ff., 53, 55 f., 62 ff., 67, 84 f., 89, 95, 99 ff., 104, 107, 111 f., 153, 154

Caracciolo, R. 76, 79 ff., 89, 152
Carcaterra 62
Carrasco, A. 39
Carrió, G. R. 3, 5, 37
Cobo del Rosal, M. 122 f., 126
Conte, A. 61, 66

Davis, D. H. 60
De Otto, I. 142 f.
Del Vecchio, G. 3
Díez-Picazo, L. 14,
Dworkin, R. 1 ff., 9, 14, 28, 37, 52, 141

Echave, D. T. 91, 102, 111
Esser, J. 3

Fishkin, J. S. 128
Franco, F. 146

García de Enterría, E. 3, 143 ff., 150
Gil Robles, G. 64
González Lagier, D. 58, 81, 94, 111
Greenawalt, K. 147

Guastini, R.o 3, 62, 74, 93
Guibourg, R. 91, 102, 111, 152, 156
Gullón, A. 14

Hare, Richard M. 130 ff.
Hart, H. L. A. 1, 6, 12, 27, 44 f., 54 f., 61, 66 f., 141, 146 f., 158 f., 161, 167
Hernández Marín, R. 53 f., 56, 99
Holmes, St. 60, 66, 85

Jescheck 122
Jori, M. 73 f.

Kant, I. 108 f., 167, 169
Kelsen, H. 49 ff., 65, 133, 141, 146

Laporta, F. J. 116 ff., 146
Luhmann, N. 13
Lukes, St. 15 f.

MacCormick, N. 46, 51 ff., 60, 85, 151
Maurach 122
Mendonça, D. 76 ff., 84 ff., 156
Mezger 122
Mir Puig, S. 121 f., 124, 133
Moreso, J. J. 76 ff., 84 ff., 155

Navarro, P. E. 76 ff., 84 ff., 155
Nino, C. S. 114, 131 f., 135, 151

Peces-Barba Martínez, G. 134, 145 f.
Peczenik, A. 26 f., 31, 34 ff., 41 ff.
Prieto Sanchís, L. 26 ff., 32, 36 ff.

Rawls, J. 15, 128 f.
Raz, J. 6, 12, 33, 36, 58, 61, 66, 88, 105, 108, 128, 144, 154, 156, 158 f., 167, 169
Redondo, C. 155
Regan, D. 33
Rescher, N. 130 ff., 135

Rodilla, M. A. 64
Ross, A. 62, 91 f., 94, 99, 112
Rubio Llorente, F. 134
Ruiz Manero, J. 28, 37 f., 42, 45, 49, 76, 116, 141, 151, 154
Ruiz Miguel, A. 146

Schauer, F. 33
Searle, J. 58, 62
Silva Sánchez, J. M. 123, 125

Summers, R. S. 24

Urquijo, M. E. 91, 102, 111

Vernengo, R. J. 116
Vives Antón, T. S. 122 f., 126

Wright, G. H. von 9, 46 ff., 77, 81 f., 92, 94 ff., 99, 111 f., 129, 135, 155

Law and Philosophy Library

1. E. Bulygin, J.-L. Gardies and I. Niiniluoto (eds.): *Man, Law and Modern Forms of Life.* With an Introduction by M.D. Bayles. 1985　　ISBN 90-277-1869-5
2. W. Sadurski: *Giving Desert Its Due.* Social Justice and Legal Theory. 1985
　　ISBN 90-277-1941-1
3. N. MacCormick and O. Weinberger: *An Institutional Theory of Law.* New Approaches to Legal Positivism. 1986　　ISBN 90-277-2079-7
4. A. Aarnio: *The Rational as Reasonable.* A Treatise on Legal Justification. 1987
　　ISBN 90-277-2276-5
5. M.D. Bayles: *Principles of Law.* A Normative Analysis. 1987
　　ISBN 90-277-2412-1; Pb: 90-277-2413-X
6. A. Soeteman: *Logic in Law.* Remarks on Logic and Rationality in Normative Reasoning, Especially in Law. 1989　　ISBN 0-7923-0042-4
7. C.T. Sistare: *Responsibility and Criminal Liability.* 1989　　ISBN 0-7923-0396-2
8. A. Peczenik: *On Law and Reason.* 1989　　ISBN 0-7923-0444-6
9. W. Sadurski: *Moral Pluralism and Legal Neutrality.* 1990　　ISBN 0-7923-0565-5
10. M.D. Bayles: *Procedural Justice.* Allocating to Individuals. 1990　　ISBN 0-7923-0567-1
11. P. Nerhot (ed.): *Law, Interpretation and Reality.* Essays in Epistemology, Hermeneutics and Jurisprudence. 1990　　ISBN 0-7923-0593-0
12. A.W. Norrie: *Law, Ideology and Punishment.* Retrieval and Critique of the Liberal Ideal of Criminal Justice. 1991　　ISBN 0-7923-1013-6
13. P. Nerhot (ed.): *Legal Knowledge and Analogy.* Fragments of Legal Epistemology, Hermeneutics and Linguistics. 1991　　ISBN 0-7923-1065-9
14. O. Weinberger: *Law, Institution and Legal Politics.* Fundamental Problems of Legal Theory and Social Philosophy. 1991　　ISBN 0-7923-1143-4
15. J. Wróblewski: *The Judicial Application of Law.* Edited by Z. Bańkowski and N. MacCormick. 1992　　ISBN 0-7923-1569-3
16. T. Wilhelmsson: *Critical Studies in Private Law.* A Treatise on Need-Rational Principles in Modern Law. 1992　　ISBN 0-7923-1659-2
17. M.D. Bayles: *Hart's Legal Philosophy.* An Examination. 1992　　ISBN 0-7923-1981-8
18. D.W.P. Ruiter: *Institutional Legal Facts.* Legal Powers and their Effects. 1993
　　ISBN 0-7923-2441-2
19. J. Schonsheck: *On Criminalization.* An Essay in the Philosophy of the Criminal Law. 1994
　　ISBN 0-7923-2663-6
20. R.P. Malloy and J. Evensky (eds.): *Adam Smith and the Philosophy of Law and Economics.* 1994　　ISBN 0-7923-2796-9
21. Z. Bankowski, I. White and U. Hahn (eds.): *Informatics and the Foundations of Legal Reasoning.* 1995　　ISBN 0-7923-3455-8
22. E. Lagerspetz: *The Opposite Mirrors.* An Essay on the Conventionalist Theory of Institutions. 1995　　ISBN 0-7923-3325-X
23. M. van Hees: *Rights and Decisions.* Formal Models of Law and Liberalism. 1995
　　ISBN 0-7923-3754-9
24. B. Anderson: *"Discovery" in Legal Decision-Making.* 1996　　ISBN 0-7923-3981-9

Law and Philosophy Library

25. S. Urbina: *Reason, Democracy, Society.* A Study on the Basis of Legal Thinking. 1996
    ISBN 0-7923-4262-3
26. E. Attwooll: *The Tapestry of the Law.* Scotland, Legal Culture and Legal Theory. 1997
    ISBN 0-7923-4310-7
27. J.C. Hage: *Reasoning with Rules.* An Essay on Legal Reasoning and Its Underlying Logic. 1997
    ISBN 0-7923-4325-5
28. R.A. Hillman: *The Richness of Contract Law.* An Analysis and Critique of Contemporary Theories of Contract Law. 1997
    ISBN 0-7923-4336-0
29. C. Wellman: *An Approach to Rights.* Studies in the Philosophy of Law and Morals. 1997
    ISBN 0-7923-4467-7
30. B. van Roermund: *Law, Narrative and Reality.* An Essay in Intercepting Politics. 1997
    ISBN 0-7923-4621-1
31. I. Ward: *Kantianism, Postmodernism and Critical Legal Thought.* 1997
    ISBN 0-7923-4745-5
32. H. Prakken: *Logical Tools for Modelling Legal Argument.* A Study of Defeasible Reasoning in Law. 1997
    ISBN 0-7923-4776-5
33. T. May: *Autonomy, Authority and Moral Responsibility.* 1998   ISBN 0-7923-4851-6
34. M. Atienza and J.R. Manero: *A Theory of Legal Sentences.* 1997   ISBN 0-7923-4856-7

KLUWER ACADEMIC PUBLISHERS – DORDRECHT / BOSTON / LONDON